T0214240

Lecture Notes in Bioinformatics 12508

Subseries of Lecture Notes in Computer Science

More information about this subseries at http://www.springer.com/series/5381

George Bebis · Max Alekseyev ·
Heyrim Cho · Jana Gevertz ·
Maria Rodriguez Martinez (Eds.)

Mathematical and Computational Oncology

Second International Symposium, ISMCO 2020
San Diego, CA, USA, October 8–10, 2020
Proceedings

 Springer

Editors
George Bebis
University of Nevada Reno
Reno, NV, USA

Max Alekseyev
George Washington University
Ashburn, VA, USA

Heyrim Cho
University of California, Riverside
Riverside, CA, USA

Jana Gevertz
The College of New Jersey
Ewing, NJ, USA

Maria Rodriguez Martinez
IBM Research - Zurich
Rüschlikon, Switzerland

ISSN 0302-9743   ISSN 1611-3349   (electronic)
Lecture Notes in Bioinformatics
ISBN 978-3-030-64510-6   ISBN 978-3-030-64511-3   (eBook)
https://doi.org/10.1007/978-3-030-64511-3

LNCS Sublibrary: SL8 – Bioinformatics

This Springer imprint is published by the registered company Springer Nature Switzerland AG
The registered company address is: Gewerbestrasse 11, 6330 Cham, Switzerland

Preface

It is with great pleasure that we welcome you to the proceedings of the 2nd International Symposium on Mathematical and Computational Oncology (ISMCO 2020), which was held virtually during October 8–10, 2020.

Despite significant advances in the understanding of the principal mechanisms leading to various cancer types, less progress has been made toward developing patient-specific treatments. Advanced mathematical and computational models could play a significant role in examining the most effective patient-specific therapies. The purpose of ISMCO is to provide a common interdisciplinary forum for mathematicians, scientists, engineers, and clinical oncologists throughout the world to present and discuss their latest research findings, ideas, developments, and applications in mathematical and computational oncology. In particular, ISMCO aspires to enable the forging of stronger relationships among mathematics and physical sciences, computer science, data science, engineering, and oncology with the goal of developing new insights into the pathogenesis and treatment of malignancies.

The program included six keynote presentations, five oral sessions, two poster sessions, two panel discussions, one tutorial, and one workshop. We received 28 submissions from which we accepted 12 submissions for oral presentation and 10 submissions for poster presentations. This LNBI volume includes only the full/short papers accepted for presentation; all abstracts accepted for presentation appeared in an online volume which was published by Frontiers (link is provided on the ISMCO website).

All submissions were reviewed with an emphasis on the potential to contribute to the state of the art in the field. Selection criteria included accuracy and originality of ideas, clarity and significance of results, and presentation quality. The review process was quite rigorous, involving at least three independent double-blind reviews followed by several days of discussion. During the discussion period we tried to correct anomalies and errors that might have existed in the initial reviews. Despite our efforts, we recognize that some papers worthy of inclusion may not have been included in the program. We offer our sincere apologies to authors whose contributions might have been overlooked.

Many contributed to the success of ISMCO 2020. First and foremost, we are grateful to the Steering, Organizing, and Program Committees; they strongly welcomed, supported, and promoted the organization of this new meeting. Second, we are deeply indebted to the keynote speakers who warmly accepted our invitation to talk at ISMCO 2020; their reputation in mathematical and computational oncology added significant value and excitement to the meeting. Next, we wish to thank the authors who submitted their work to ISMCO 2020 and the reviewers who helped us to evaluate the quality of the submissions. It was because of their contributions that we succeeded in putting together a technical program of high quality. Finally, we would like to express our appreciation to Springer, Frontiers, and the International Society for Computational Biology (ISCB) for supporting ISMCO 2020.

We sincerely hope that despite the difficulties due to the pandemic, ISMCO 2020 offered participants opportunities for professional growth. We look forward to many more successful meetings in mathematical and computational oncology.

October 2020

George Bebis
Max Alekseyev
Heyrim Cho
Jana Gevertz
Maria Rodriguez Martinez

Organization

Steering Committee

Anastasiadis Panagiotis	Mayo Clinic, USA
Bebis George (Chair)	University of Nevada, Reno, USA
Jackson Trachette	University of Michigan, USA
Levy Doron	University of Maryland, USA
Rockne Russell	Beckman Research Institute at City of Hope, USA
Vasmatzis George	Mayo Clinic, USA
Yankeelov Thomas	The University of Texas at Austin, USA

Program Chairs

Alekseyev Max	George Washington University, USA
Cho Heyrim	University of California, Riverside, USA
Gevertz Jana	The College of New Jersey, USA
Martinez Maria Rodriguez	IBM, Zurich Research Laboratory, Switzerland

Publicity Chairs

Loss Leandro	QuantaVerse, USA
Petereit Juli	University of Nevada, Reno, USA

Tutorials and Special Tracks Chairs

Nguyen Tin	University of Nevada, Reno, USA
Scalzo Fabien	University of California, Los Angeles, USA

Web Master

Isayas Berhe Adhanom	University of Nevada, Reno, USA

Program Committee

Pankaj Agarwal	BioInfi, USA
Max Alekseyev	George Washington University, USA
Dinler Antunes	Rice University, USA
Erman Ayday	Case Western Reserve University, USA, and Bilkent University, Turkey
Matteo Barberis	University of Surrey, UK
Ali Bashashati	The University of British Columbia and BC Cancer Agency, Canada

George Bebis	University of Nevada, Reno, USA
Takis Benos	University of Pittsburgh, USA
Debswapna Bhattacharya	Auburn University, USA
Mary Regina Boland	University of Pennsylvania, USA
Ernesto Augusto Bueno Da Fonseca Lima	The University of Texas at Austin, USA
Anton Buzdin	Omicsway Corp., USA
Raffaele Calogero	University of Torino, Italy
Hannah Carter	University of California, San Diego, USA
Young-Hwan Chang	Oregon Health and Science University, USA
Aristotelis Chatziioannou	National Hellenic Research Foundation, Greece
Jake Chen	University of Alabama at Birmingham, USA
Ken Chen	The University of Texas MD Anderson Cancer Center, USA
Luonan Chen	Osaka Sangyo University, Japan
You Chen	Vanderbilt University, USA
Juan Carlos Chimal	IPN - Computing Research Center, Mexico
Heyrim Cho	University of California, Riverside, USA
Giovanni Ciriello	University of Lausanne, Switzerland
Jean Clairambault	Inria, France
Francois Cornelis	Sorbonne University, France
James Costello	University of Colorado Anschutz Medical Campus, USA
Paul-Henry Cournede	CentraleSupélec, France
Morgan Craig	University of Montreal, Canada
Simona Cristea	Harvard University, USA
Kit Curtius	University of California, San Diego, USA
Francesca Demichelis	University of Trento, Italy
Anne Deslattes Mays	Science and Technology Consulting, LLC, USA
Mohammed El-Kebir	University of Illinois at Urbana-Champaign, USA
Peter Elkin	Ontolimatics, USA
Sally Ellingson	University of Kentucky, USA
Funda Ergun	Indiana University Bloomington, USA
Alexander Fletcher	The University of Sheffield, UK
Anthony Fodor	University of North Carolina at Charlotte, USA
Terry Gaasterland	University of California, San Diego, USA
Andrew Gentles	Stanford University, USA
Jana Gevertz	The College of New Jersey, USA
Preetam Ghosh	Virginia Commonwealth University, USA
Jeremy Goecks	Oregon Health and Science University, USA
Thierry Goudon	Inria, France
David Robert Grimes	Dublin City University, Ireland
Wei Gu	University of Luxembourg, Luxembourg
Hiroshi Haeno	National Cancer Center Hospital, Japan
Michael Hallett	Concordia University, Canada

Arif Harmanci	The University of Texas Health Science Center at Houston, USA
Leonard Harris	University of Arkansas, USA
Andrea Hawkins-Daarud	Mayo Clinic, USA
Harry Hochheiser	University of Pittsburgh, USA
Vasant Honavar	The Pennsylvania State University, USA
David Hormuth	The University of Texas at Austin, USA
Florence Hubert	Aix-Marseille University, France
Trachette Jackson	University of Michigan, USA
Katharina Jahn	ETH Zurich, Switzerland
Harsh Jain	Florida State University, USA
Peter Jeavons	University of Oxford, UK
Mamoru Kato	National Cancer Center Hospital, Japan
Ioannis Kavakiotis	University of Thessaly, Greece
Artem Kaznatcheev	University of Oxford, UK
Seungchan Kim	Prairie View A&M University, USA
Marek Kimmel	Rice University, USA
Maria Klapa	Foundation for Research and Technology - Hellas, Greece
Rachel Kolodny	Harvard Medical School, USA
Arjun Krishnan	Michigan State University, USA
Ellen Kuhl	Stanford University, USA
Parvesh Kumar	University of Nevada, Las Vegas, USA
Nguyen Quoc Khanh Le	Nanyang Technological University, Singapore
Hayan Lee	Stanford University, USA
Richard Levenson	University of California, Davis, USA
Doron Levy	University of Maryland, USA
Xiaotong Li	Yale University, USA
Michal Linial	The Hebrew University of Jerusalem, Israel
Jiajian Liu	Merck-KGaA, Germany
Zhandong Liu	Baylor College of Medicine, USA
Guillermo Lorenzo	The University of Texas at Austin, USA, and The University of Pavia, Italy
Leandro Loss	QuantaVerse, ITU, USA, and ESSCA, France
Shaoke Lou	Yale University, USA
Adam MacLean	University of Southern California, USA
Sokratis Makrogiannis	Delaware State University, USA
Brad Malin	Vanderbilt University, USA
Kostas Marias	Foundation for Research and Technology - Hellas, Greece
Scott Markel	Dassault Systèmes BIOVIA, France
Peter McCaffrey	University of Texas Medical Branch, USA
Jörg Menche	University of Vienna, Austria
Abu Mosa	University of Missouri, USA
Neha Murad	Calico Labs, USA

Rick Stevens	The University of Chicago and Argonne National Laboratory, USA
Peter Sykacek	University of Natural Resources and Life Sciences, Austria
Ahmad Tafti	Mayo Clinic, USA
Haixu Tang	Indiana University Bloomington, USA
Umit Topaloglu	Wake Forest University, USA
Tamir Tuller	Tel Aviv University, Israel
Jack Tuszynski	University of Alberta, Canada
Daniela Ushizima	Lawrence Berkeley National Laboratory, USA
George Vasmatzis	Mayo Clinic, USA
Eduardo Vilar-Sanchez	The University of Texas MD Anderson Cancer Center, USA
Yannick Viossat	Université Paris-Dauphine, France
Li-San Wang	University of Pennsylvania, USA
Yanshan Wang	Mayo Clinic, USA
Mark Wass	University of Kent, UK
Matt Williams	Imperial College London, UK
Yanji Xu	NIH, USA
Rui Yamaguchi	Aichi Cancer Center Research Institute, Japan
Nikos Zacharakis	NCI NIH, USA
Meirav Zehavi	Ben-Gurion University of the Negev, Israel
Alex Zelikovsky	Georgia State University, USA
Maryam Zolnoori	Mayo Clinic, USA

Additional Reviewers

Bose, Priyankar
Bottcher, Rene
Huang, Yuefan
Liang, Shaoheng

Liu, Liang
Su, Jing
Wang, Yuanxin
Yao, Jin

Abstracts of Keynote Talks

Fighting Drug Resistance with Math

Doron Levy

University of Maryland at College Park, USA

Abstract. The emergence of drug-resistance is a major challenge in chemotherapy. In this talk we will overview some of our recent mathematical models for describing the dynamics of drug-resistance in solid tumors. These models follow the dynamics of the tumor, assuming that the cancer cell population depends on a phenotype variable that corresponds to the resistance level to a cytotoxic drug. Under certain conditions, our models predict that multiple resistant traits emerge at different locations within the tumor, corresponding to heterogeneous tumors. We show that a higher drug dosage may delay a relapse, yet, when this happens, a more resistant trait emerges. We will show how mathematics can be used to propose an efficient drug schedule aiming at minimizing the growth rate of the most resistant trait, and how such resistant cells can be eliminated.

To Function or Not to Function

Sridhar Hannenhalli

National Cancer Institute, USA

Abstract. The functions of only a minority of genes in any species is known. And even in those cases the functional annotation is highly incomplete and largely devoid of context. At an even more fundamental level, how can we know whether a gene serves any relevant biological function in a given context? In this informal presentation we will discuss a few vignettes related to the broad questions of context-specific functions of genes, in a variety of contexts from bacterial response to drugs, normal tissues, and cancer.

From Mathematical Modelling of Cancer Cell Plasticity to Philosophy of Cancer

Jean Clairambault

Inria and Sorbonne University, France

Abstract. In this talk, I will suggest that cancer is fundamentally a disease of the control of cell differentiation in multicellular organisms, uncontrolled cell proliferation being a mere consequence of blockade, or unbalance, of cell differentiations. Cancer cell populations, that can reverse the sense of differentiations, are extremely plastic and able to adapt without mutations their phenotypes to transiently resist drug insults, which is likely due to the reactivation of ancient, normally silenced, genes. Stepping from mathematical models of non genetic plasticity in cancer cell populations and questions they raise, I will propose an evolutionary biology approach to shed light on this problem both from a theoretical viewpoint by a description of multicellular organisms in terms of multi-level structures, which integrate function and matter from lower to upper levels, and from a practical point of view oriented towards cancer therapeutics, as cancer is primarily a failure of multicellularity in animals and humans. This approach resorts to the emergent field of knowledge named philosophy of cancer.

Quantitative Molecular Dissection of Cancer Evolution

Ken Chen

MD Anderson Cancer Center, USA

Abstract. A cancer initiates, grows, and metastasizes over time and space. It often involves dynamic, genotypical and phenotypical evolution and interaction of millions of cells, belonging to hundreds of cell types. Successful cancer prevention and treatment require quantitative approaches that can identify key factors that are causal to cancer evolution and can be therapeutically intervened. Achieving such a goal has been challenging, due partly to limitations in data collection, analysis, and interpretation. In this talk, I will highlight ongoing efforts that involve various aspects of experimental design, application of high-throughput multiomics technologies such as single-cell DNA, RNA, and ATAC sequencing, and statistical computational approaches to tackle such an important challenge.

Deep Learning for Clinically Actionable Cancer Pathology Feature Detection

Alexander Pearson

UChicago Medicine, USA

Abstract. An immense amount of information is stored in the spatial information of tumor histology. Access to contemporary neural network architecture, the decreasing cost of hardware, and the immense amount of available pathologic and genomic data have contributed to rapid innovation in digital pathology analysis. We will describe the basics of deep learning and how it can readily be applied to histology data. We will further describe specific applications to clinical oncology, including defining molecular treatment targets directly from histology. Finally we will discuss the state of the field, and barriers and opportunities for wide deployment of deep learning in our healthcare system.

Enriching Cancer Research Through Unconventional Collaborations

Jennifer Couch

National Cancer Institute, USA

Abstract. Cancer is complicated and complex, arising in multiple cell types and tissue of origin, initiating and progressing differently in different individuals or under different conditions, with effects crossing multiple biological scales. It is also adaptive, evolving at both the molecular and cellular scales during treatment. Because of this, cancer researchers are often early adopters of new technologies, methods, and approaches and they routinely adopt tools and methods originally developed in other, very different fields for use in modeling, understanding, and combating cancer. Systems biology, with its combination of experimental biology and mathematical modeling, plays an increasingly important role in cancer research. Advances in data science, high performance computing, and artificial intelligence have led to increasingly creative adoption of these tools and approaches in cancer research. But high-dimensional data, complicated problems, and collaborative problem solving also exist in areas such as entertainment video games and virtual and augmented reality. And cancer data and its contextual information can come from a variety of sources including patients and the public through citizen science and crowdsourcing. This talk will explore this broader collaborative and creative space and high-light both on-going and new programs supported by the National Cancer Institute to create opportunities for cross-field interactions and to accelerate new collaborations and new inter-disciplinary approaches to cancer research.

Contents

General Cancer Computational Biology

Posters

Invited Talk

Plasticity in Cancer Cell Populations: Biology, Mathematics and Philosophy of Cancer

Jean Clairambault[✉][iD]

Inria and LJLL, Sorbonne University, Paris, France
jean.clairambault@inria.fr,
https://who.rocq.inria.fr/Jean.Clairambault/Jean_Clairambault_en.html

Abstract. In this presentation that partly subsumes and summarises in the form of adapted excerpts some recent articles of which I am author or co-author [2,3,11], I suggest that cancer is fundamentally a disease of the control of cell differentiation in multicellular organisms, uncontrolled cell proliferation being a mere consequence of blockade, or unbalance, of cell differentiations. Cancer cell populations, that can reverse the sense of differentiations, are extremely plastic and able to adapt without mutations their phenotypes in order to transiently resist drug insults [10], which is likely due to the reactivation of ancient, normally silenced, genes [5,7,12]. Stepping from *mathematical models of non genetic plasticity* in cancer cell populations [4,6] and questions they raise, I propose an *evolutionary biology* approach to shed light on this problem a) from a theoretical viewpoint by a description of multicellular organisms in terms of multi-level structures, which integrate function and matter from lower to upper levels, and b) from a practical point of view by proposing future tracks for cancer therapeutics, as cancer is primarily a failure of multicellularity in animals and humans. This approach resorts to the emergent field of knowledge known as *philosophy of cancer* [1,8,9].

Keywords: Cancer cells · Plasticity · Mathematical models · Philosophy of cancer

1 Introduction

Coherent multicellular organisms are not only cohesive from a spatial, anatomical point of view, but also coherent from the phenotypic and cell-functional point of view of compatibility, cooperativity (division of tasks) between cells and tissues, making possible the achievement of a stable, functional and reproductive whole.

I make here the simple hypothesis of a system of communication ways between trees of differentiation, relying on the control of transcription factors

---

Invited paper.

© Springer Nature Switzerland AG 2020
G. Bebis et al. (Eds.): ISMCO 2020, LNBI 12508, pp. 3–9, 2020.
https://doi.org/10.1007/978-3-030-64511-3_1

that determine differentiations, and that I call "the cohesion watch". It may be considered as a part of the immune system, whose armed force is the immune response, innate as well as adaptive, humoral and cellular, but is not the whole of the immune system, that I view as the conductor of the unity of the organism. Within the immune system in this extended vision that is thus more general than the immune response, this "cohesion watch" is in charge of the control of compatibilities and cooperations between the anatomical and the phenotypic/cell-functional systems, ultimately leading to an anatomically cohesive and functionally coherent multicellular organism.

2 Plasticity in Cancer Cell Populations

2.1 Non Genetic Phenotype Switching

Phenotype switching, or more generally continuous phenotype-determined cell plasticity, is an essential process originally observed during development, but is also now recognised as an important phenomenon upon injury and disease. One of the best described examples of phenotypic switching in cancer depends on the process of epithelial to mesenchymal transition (EMT) and its reversion (MET) [11]. In the context of anti-cancer therapies, cell plasticity enables tumour cells to change to a cell phenotypic identity that may be dependent or not on the drug target, without additional secondary genetic mutations. Indeed, the discovery of oncogenic-driven mutations favoured the development of diverse targeted therapies and showed unprecedented clinical response. Unfortunately, responses are in general incomplete and transient, as resistances develop upon continuous treatment exposure. Along with well-known genetic alterations, cell plasticity has recently emerged as an unavoidable contributor to therapy evasion.

2.2 Transient Drug-Induced Tolerance in Cancer

The biological mechanisms that induce cancer cell plasticity upon drug treatment remain to be fully established. Nonetheless, they seem to involve a transition whereby tumour cells undergo a slow proliferating drug tolerant state, nailed the name of drug-tolerant persisters (DTPs), before possibly further developing secondary mutational drug resistance. Persisters were firstly described in bacteria upon antibiotic challenges. Similarly, a sub-population of non-small cell lung cancer (NSCLC) cells has been shown to engage in a reversible phenotypic change in which DTPs survive the initial onslaught of anti-cancer therapies [10]. Similar phenomena were observed in glioblastoma and melanoma. Biological observations support the idea that a tiny subpopulation, diverging from predominant cell phenotype prior to targeted therapy, may be subjected to a combination of Lamarckian plasticity and Darwinian selection upon anti-cancer therapies.

2.3 Dedifferentiation and Transdifferentiation

Conversion of lineage has been extensively studied in the context of development. The well-known, metaphoric, Waddington landscape has been proposed to illustrate the fact that a progenitor cell normally rolls down within epigenetic differentiation valleys and can develop, due to phenotypic bifurcations, into the various terminally differentiated tissue types that constitute a coherent multicellular organism (see also Sect. 3). In the context of cancer, dedifferentiation and transdifferentiation have been observed upon therapeutic challenges, which suggests a possible plasticity in the cancers Waddington landscape, with metaphorically flattened valleys and lowered epigenetic barriers.

3 Mathematics of Plasticity with Therapeutic Control

3.1 How to Mathematically Model Plasticity in Cancer

Cell plasticity is the ability of cells to change their phenotypes without genetic mutations in response to environmental cues. In a series of papers starting in 2013, a team of mathematicians, to which I belong, at Laboratoire Jacques-Louis Lions, Sorbonne University, Paris, taking plasticity to be a concept of the continuum, and initially stimulated by an article published in 2010 [10] that reported reversible resistance in a cancer cell culture exposed to massive doses of drugs, gradually and completely reversed when the drug was withdrawn from the culture, tackled the question of understanding and predicting this plastic behaviour of cancer cell populations by mathematical modelling (reviewed in [4]). The behaviour of these *plastic* cell populations was modelled by partial differential equations (PDEs) in which the structuring variable, i.e., the parameter-like one that codes for the biological variability of interest, i.e., the *heterogeneity* of cell population, was here a continuous variable representing the expression of a continuous resistance phenotype, from 0 (totally sensitive) to 1 (totally resistant). The modelled cell populations are able to change their phenotypes under high drug pressure so as to become *reversibly* resistant to the drug, meaning by this the fact that, as real cells when the drug was rinsed from the culture, the cell population spontaneously became sensitive again to the same drug [10].

3.2 Adaptive Dynamics: Asymptotic Behaviour of Cell Populations

Such mathematical models, that resort to the field of adaptive dynamics [4,6] consist of partial differential equations (PDEs) structured in continuous phenotypes coding for the expression of drug resistance genes; they involve different functions representing targets for different drugs, cytotoxic and cytostatic, with complementary effects in limiting tumour growth. These phenotypes evolve continuously under drug exposure, and their fate governs the evolution of the cell population under treatment.

This evolutionary point of view, which relies on biological observations and resulting modelling assumptions, naturally extends to questioning the very nature of cancer as evolutionary disease, seen not only at the short time scale of a human life, but also at the billion year-long time scale of Darwinian evolution, from unicellular organisms to evolved multicellular organs such as animals and man. Such questioning, not so recent, but recently revived [5,7,12] in cancer studies, may have consequences for understanding and treating cancer.

3.3 Theoretical Therapeutics: Multi-targeted Optimal Control

These mathematical models, intended to represent the effects of a cancer treatment on cell populations, and ultimately on patients, with the aim to overcome their capacities of resistance induced by the treatment itself, naturally gave later rise to the proposal of theoretically optimised therapeutic strategies. Methods of optimal control have been used, taking inevitable emergence of drug resistance into account, to achieve the best strategies to contain the expansion of a tumour. Such strategies, that have recently been the object of active research aim at containing or eradicating cancer growth, avoiding the two major pitfalls of treatments in clinical oncology, namely unwanted toxic side effects in healthy cell populations and emergence of resistance in cancer populations [4,6].

4 Evolutionary Biology and Philosophy of Cancer

4.1 'Nothing Makes Sense in Biology Except in the Light of Evolution'

This celebrated motto of the zoologist Theodosius Dobzhansky is particularly true in developmental biology and in cancer, that is firstly an evolutionary disease, more precisely a disease of evolutionary multicellular organisation because it may be considered as a backward step in the course of evolution towards organised multicellularity, according to the atavistic theory of cancer ([5,12], see next Subsection). Cancer thus represents an evolution at the time scale of a living multicellular organism, anatomically localised in a given organ, possibly extended to other tissues by remote metastases, towards genetically new unicellular species developing, likely with branching by successions of mutations, their diversity at the expense of the host organism. I contend that the disappearance of successive physiological control mechanisms puts cancer cell populations in the state of a very primitive multicellular organisation, as proposed below.

4.2 The Atavistic Theory of Cancer

In 2011, independently, physicists Paul Davies and Charles Lineweaver, and oncologist Mark Vincent avocated the idea that cancer is a de-repression of a default survival program common to all cells [12] which was expressed by Davies and Lineweaver as the atavistic theory of cancer [5]. Alternatively saying, cancer is a disease of the evolution of multicellular organisms in which a localised collection of cells organises itself and proliferates for its own benefit (possibly trying to reinvent the wheel of multicellularity and bound to failure). This hypothesis has been assessed from phylostratigraphic analyses of the genomes of different species that allowed to establish links between the genes that are essential to multicellularity and those that are altered in cancer, and from more recent studies that elicit disrupted relationships between genes of multicellularity and genes that are disrupted in cancer. This emerging field of research of course presupposes that cancer is an evolutionary disease of multicellular organisms, 'evolutionary' here meaning related to the Darwinian evolution of living species. In this respect, plasticity is related to loss of control of the differentiations that make a multicellular organism coherent and functional, and any disruption in this control may lead to cancer, without necessarily resorting to cancer stem cells. In other words, loss of control on differentiation, which is most likely related to defective control by the immune system, *is actually the plasticity of cancer.*

4.3   Failed Control of Differentiations: Cancer is a Failure of Cohesion

The task of the hypothesised cohesion watch, part of an extended version of the immune system, is thus to ensure compatibilities a) between morphogens of the body plan, able to drive it actually from the zygote to a constituted multicellular being in an irreversible way within the 3D space of cells of a given individual (in jawed vertebrates defined by the major hiostocompatibility complex, MHC, and by some likely equivalent forerunners in non-vertebrates); b) between phenotypic functionalities, ensuring compatibility between differentiation trees that yield lineages within a given subpopulation during embryogenesis, and ultimately between cooperating subpopulations (division of work) of terminally differentiated cells; c) between the body plan space distribution and the time distribution of phenotypes in each epigenetic landscape attached to the body plan.

To mentally illustrate this construction, I propose as a further metaphor the wickerwork basket. Starting from a circle endowed with lots of connections between its elements, that is supposed to represent the body plan, functional willow-like twigs stem from each of these elements, representing the great physiological functions of the organism. If no weaving is made between these twigs, the whole set will consist of just flexible differentiation functionalities of a family of cell types, floating freely in the surrounding space, unrelated to each other. No cohesion, no division of labour can result from such unwoven twigs and trees. The task of the cohesion watch is to ensure such weaving during development, along the twigs from stem cells until tips that are terminally differentiated cells. This naturally includes the solidity of the willow twigs (breaches along the vertical axis resulting in blocked differentiations), but the main part of the cohesion watch is to ensure compatibility between neighbouring twigs.

4.4   Speculations on the Possible Future of Cancer Therapeutics

Rather than fighting uncontrolled proliferation, could we repair altered control on differentiations? Cell-killing strategies, be they relying on chemotherapies or on modern immune cell-enhancing drugs, miss the basic targets, that are differentiation sites, and better work would undoubtedly be done by enforcing connections ensured by the cohesion watch, rather than by killing cheater cells. Or else, trying to illustrate this goal with a sociological metaphor: rather than killing cheater cells by cannonade (i.e., by chemotherapies) or by enforcing the aggressiveness of the police (i.e., by immune checkpoint inhibitors), would it not be better to imagine how to enforce natural cohesion between cells? To be able to do this, a better understanding of the mechanisms of control of differentiation at the level of local transcription factors and at the level of chromatin is needed.

5 Conclusion

Far from considerations on evolution of a cell population at the time scale of a human life - my starting point [4, 6] - , that undoubtedly present a high interest in therapeutics, I have presented in this study an evolutionary point of view on cancer in a billion-year perspective that, from questions on plasticity in cancer, drove me to develop ideas and speculations that resort to what is now known as *philosophy of cancer* [1, 8, 9], thus beginning to tread a long and winding path towards a fundamental understanding of multicellularity and of its alterations in cancer, that should lead to correct impaired control of differentiation, rather than, or at least together with, control of proliferation.

References

1. Bertolaso, M.: Philosophy of Cancer. HPTLS, vol. 18. Springer, Dordrecht (2016). https://doi.org/10.1007/978-94-024-0865-2
2. Chisholm, R.H., Lorenzi, T., Clairambault, J.: Cell population heterogeneity and evolution towards drug resistance in cancer: biological and mathematical assessment, theoretical treatment optimisation. Biochim. Biophys. Acta **1860**, 2627–2645 (2016). https://doi.org/10.1016/j.bbagen.2016.06.009
3. Clairambault, J.: Stepping from modelling cancer plasticity to philosophy of cancer. Front. Genet. **11**, 1160 (2020). https://doi.org/10.3389/fgene.2020.579738
4. Clairambault, J., Pouchol, C.: A survey of adaptive cell population dynamics models of emergence of drug resistance in cancer, and open questions about evolution and cancer. Biomath **8**, 23 (2019). https://doi.org/10.11145/j.biomath.2019.05.147. 1905147
5. Davies, P.C.W., Lineweaver, C.H.: Cancer tumors as Metazoa 1.0: tapping genes of ancient ancestors. Phys. Biol. **8**(1), 015001–015007 (2011). https://doi.org/10.1088/1478-3975/8/1/015001
6. Pouchol, C., Clairambault, J., Lorz, A., Trélat, E.: Asymptotic analysis and optimal control of an integro-differential system modelling healthy and cancer cells exposed to chemotherapy. J. Math. Pures Appl. **116**, 268–308 (2018). https://doi.org/10.1016/j.matpur.2017.10.007
7. Lineweaver, C.H., Davies, P.C.W., Vincent, M.: Targeting cancers weaknesses (not its strengths): therapeutic strategies suggested by the atavistic model. BioEssays **36**, 827–835 (2014). https://doi.org/10.1002/bies.201400070

8. Pradeu, T.: The Limits of the Self. Oxford University Press, Oxford (2012)
9. Pradeu, T.: Philosophy of Immunology. Cambridge University Press, Cambridge (2019)
10. Sharma, S.V., et al.: A chromatin-mediated reversible drug-tolerant state in cancer cell subpopulations. Cell **141**, 69–80 (2010). https://doi.org/10.1016/j.cell.2010.02.027
11. Shen, S., Clairambault, J.: Cell plasticity in cancer cell populations (review) [version 1; peer review: 2 approved]. F1000Research 2020 **9**(F1000 Faculty Rev), 635–650 (2020). https://doi.org/10.12688/f1000research.24803.1
12. Vincent, M.D.: Cancer: a de-repression of a default survival program common to all cells?: a life-history perspective on the nature of cancer. Bioessays **34**(1), 72–82 (2011). https://doi.org/10.1002/bies.201100049

Statistical and Machine Learning
Methods for Cancer Research

CHIMERA: Combining Mechanistic Models and Machine Learning for Personalized Chemotherapy and Surgery Sequencing in Breast Cancer

Cristian Axenie[1,2]([✉]) and Daria Kurz[3]

[1] Audi Konfuzius-Institut Ingolstadt Lab, Ingolstadt, Germany
cristian.axenie@audi-konfuzius-institut-ingolstadt.de
[2] Technische Hochschule Ingolstadt, Esplanade 10, 85049 Ingolstadt, Germany
[3] Interdisziplinäres Brustzentrum, Helios Klinikum München West,
Steinerweg 5, 81241 Munich, Germany
daria.kurz@helios-gesundheit.de

Abstract. Mathematical and computational oncology has increased the pace of cancer research towards the advancement of personalized therapy. Serving the pressing need to exploit the currently underutilized data, such approaches bring a significant clinical advantage in tailoring the therapy. CHIMERA is a novel system that combines mechanistic modelling and machine learning for personalized chemotherapy and surgery sequencing in breast cancer. It optimizes decision-making in personalized breast cancer therapy by connecting tumor growth behaviour and chemotherapy effects through predictive modelling and learning. We demonstrate the capabilities of CHIMERA in learning simultaneously the tumor growth patterns, across several types of breast cancer, and the pharmacokinetics of a typical breast cancer chemotoxic drug. The learnt functions are subsequently used to predict how to sequence the intervention. We demonstrate the versatility of CHIMERA in simultaneously learning tumor growth and pharmacokinetics under two, typically used, chemotherapy protocol hypotheses.

Keywords: Machine learning · Chemotherapy sequencing · Breast cancer · Personalized medicine · Mathematical oncology

1 Background

As the last decades have shown, early diagnosis and new drugs have led to impressive increases in survival rates of breast cancer patients. Yet, tailoring standard treatment schemes to patient needs is still a sought for objective. A personalised approach, requires new methods that exploit tumor biology and the effect chemotoxic drugs have upon the tumor, in order to sequence the interventions [21]. Neoadjuvant therapy has grown into a well-established, safe and

© Springer Nature Switzerland AG 2020
G. Bebis et al. (Eds.): ISMCO 2020, LNBI 12508, pp. 13–24, 2020.
https://doi.org/10.1007/978-3-030-64511-3_2

often beneficial approach to breast cancer treatment. In terms of survival and overall disease progression, neoadjuvant and adjuvant treatments tend to be similar treatment choices for breast cancer [10]. Yet, the neoadjuvant treatment increases breast-conserving surgery levels and improves resectability by reducing the primary tumour. Moreover, it can support the early evaluation of the efficacy of the therapy chosen [11]. This assessment may allow the clinician to discontinue ineffective treatment or may help switch to another regimen to maximise response [1]. Patients who do not achieve a pathologic complete response after neoadjuvant chemotherapy, consider the use of adjuvant scheme [12]. Overall, it is reported that chemotherapy use in the neoadjuvant and adjuvant settings generally provides the same long-term outcome. But what is the best course of action for a particular patient? This question targets those quantifiable patient-specific factors (e.g. tumor growth curve and chemotherapy effect parameters, such as drug pharmacokinetics) that influence the sequencing of chemotherapy and surgery and taps into personalized therapy.

1.1   Formalizing Therapy Sequencing

A model for personalized sequencing should include tumor cells growth and the effects of chemotherapy and surgery under cell-kill hypotheses. This hypothesis proposes that actions of chemotoxic drugs follow first order kinetics: a given dose kills a constant proportion of a tumor cell population (rather than a constant number of cells) [20]. Assuming that the tumor size at time $t_0 = 0$ is V_0, there are two possible sequences:

- **Adjuvant chemotherapy** At time $t_0 > 0$ a fraction of the tumor is removed through surgery and subsequently chemotherapy is administered with a killing rate of $1 - e^{-k_s}$ where k_s is a rate constant. The final size after the intervention, at $t_f > t_0$ is V_{adj}.
- **Neoadjuvant chemotherapy** At time $t_0 > 0$ chemotherapy is administered with a predefined killing rate. At $t_f > t_0$ a fraction $1 - e^{-k_s}$ of the tumor is removed through surgery for a final size after the intervention V_{neoadj}.

The question of interest in our study is if $V_{adj} > V_{neoadj}$? If we consider $f(V)$ the tumor growth model and $P(t, V)$ the pharmacokinetics of the chemotherapeutic drug, we can formalize the two sequences as following:

- Sequence 1: Adjuvant setting, where size before surgery is $\frac{dv_1}{dt} = f(v_1)$, $v_1(0) = V_0, t \in [0, t_0]$ and size after surgery is $\frac{dV_1}{dt} = f(V_1) - P(t, V_1), V_1(t_0) = e^{-k_s}v_1(t_0), t \in [t_0, t_f]$. In this case, the final volume of the tumor is $V_{adj} = V_1(t_f)$.
- Sequence 2: Neoadjuvant setting, where the size before chemotherapy onset is $\frac{dv_2}{dt} = f(v_2), v_2(0) = V_0, t \in [0, t_0]$ and the size after chemotherapy onset is $\frac{dV_2}{dt} = f(V_2) - P(t, V_2), V_2(t_0) = v_2(t_0), t \in [t_0, t_f]$ respectively. Hence, for the neoadjuvant sequence, the final volume of the tumor is $V_{neoadj} = e^{-k_s}V_2(t_f)$.

Such sequencing is typically valid under some cell-kill hypotheses. The two cell-kill hypotheses we employ in our study are the log-kill hypothesis [4] and the Norton-Simon hypothesis, respectively [14]. The log-kill hypothesis states that the effect the chemotoxic drug has upon the tumor is $P(t, V) = c(t)V$, where V is the volume of the tumor or number of cells, and $c(t)$ is a function proportional with the chemotoxic drug concentration at time t. On the other side, the Norton-Simon hypothesis, defines the effect of the chemotoxic drug as $P(t, V) = c(t)f(V)$, where $f(V)$ is the tumor growth curve function depending on the volume of the tumor or number of cells. Various studies [6] considering average values over the populations of patients demonstrated that under the log-kill hypothesis $V_{adj} > V_{neoadj}$ whereas under the Norton-Simpson hypothesis $V_{adj} > V_{neoadj}$ or $V_{adj} < V_{neoadj}$. In order to ensure that such sequencing is personalized, we explore how can a machine learning algorithm extract the two functions of interest, namely tumor growth function f and pharmacokinetics effect P from data, without constraining the choice of a specific model. Such a limiting approach would be detrimental for patients as it might not capture the tumor dynamics and the effect chemotherapy has for the long-term intervention.

1.2   Models of Tumor Growth

A large variety of breast cancer tumor growth patterns were identified experimentally and clinically, and modelled over the years. Ordinary differential equations (ODE) tumor growth models [5] are typically used in cancer treatments planning. In our study, we explored three of the most representative and typically used scalar growth models, namely Logistic, von Bertalanffy, and Gompertz, described in Table 1.

Table 1. Overview of tumor growth models $f(V)$ in our study. Parameters: V - volume (or cell population size through conversion - $V(t) = V_{one\_cell} ICN e^{k_N t}$ where N is the population size, ICN is the initial cell number, V_{one_cell} is the volume of one cell and k_N is the rate constant for changes in cell number as considered in [8]), α - growth rate, β - cell death rate, λ - nutrient limited proliferation rate, k - carrying capacity of cells.

Model	Equation
Logistic [17]	$f_L(V) = \frac{dV}{dt} = \alpha V - \beta V^2$
Bertalanffy [19]	$f_B(V) = \frac{dV}{dt} = \alpha V^\lambda - \beta V$
Gompertz [5]	$f_G(V) = \frac{dV}{dt} = V(\beta - \alpha \ln V)$

1.3   Models of Chemotherapy Pharmacokinetics

Pharmacokinetic modelling is a useful tool to describe and investigate the effect of covariates in drug administration. A number of population pharmacokinetic models have described the pharmacokinetics of the taxanes drug family [22], a typical breast cancer therapy candidate. More precisely, they addressed Paclitaxel monotherapy and have provided important insight into Paclitaxel pharmacokinetics [15]. In our study, we use the data from the computational model of intracellular pharmacokinetics of Paclitaxel of [8]. The model describes the factors that determine the kinetics of Paclitaxel uptake, binding, and efflux from cells. The models demonstrated that changes in cell number were represented by changes in volume which: 1) increased with time at low initial total extracellular drug concentrations due to continued cell proliferation and 2) decreased with time at high initial total extracellular drug concentrations due to the antiproliferative and/or cytotoxic drug effects, as reported in [8]. Such nonlinear effects are patient specific and parametrizing the model needs very detailed biological specification and analysis, which in vivo might not be feasible. Another challenge regarding the clinical use of Paclitaxel is the identification of optimal treatment drug administering schedules. The difficulty is in part due to the lack of a precise understanding of individual pharmacokinetics of Paclitaxel, i.e., drug effect as a function of drug concentration and treatment duration for each patient. Such challenges motivated our study.

1.4   Objectives of the Study

The objectives of our study are to demonstrate how CHIMERA, a combination of machine learning and mechanistic modelling, can predict the chemotherapy and surgery sequencing, through:

- learning the tumor growth model $f(V)$ from tumor growth data of breast cancer, and
- learning the pharmacokinetics $P(t, V)$ of the chemotoxic dose response in the sequencing scheme,

for a truly personalized intervention in breast cancer patients.

2   Materials and Methods

In the next section we introduce the underlying mechanisms of CHIMERA as well as the experimental procedures used in our experiments.

2.1   Introducing CHIMERA

CHIMERA is an unsupervised machine learning system based on Self-Organizing Maps (SOM) [7] and Hebbian Learning (HL) [2] used for extracting underlying functional relations among correlated timeseries describing therapy variables, such as tumor growth and chemotherapy pharmacokinetics. We introduce the basic mechanisms in CHIMERA through a simple example in Fig. 1. Here, we consider data from a breast cancer growth function under sequential chemotherapy carried over 150 weeks from [3]. The two input timeseries (i.e. the number of

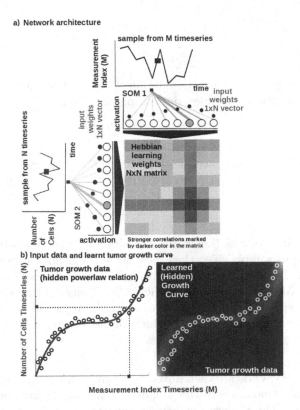

Fig. 1. Basic functionality of CHIMERA: a) basic architecture of CHIMERA: 1D SOM networks with N neurons encoding the timeseries (i.e. number of cells vs. measurement index), and a NxN Hebbian connection matrix coupling the two 1D SOMs that will eventually encode the relation between the timeseries, i.e. growth curve. b) Left: tumor growth data resembling a non-linear relation hidden in the timeseries (i.e. number of cells vs. measurement index) Data from [3]. Right: learnt tumor growth curve.

tumor cells and the irregular measurement index over the weeks) follow a cubic dependency, depicted in Fig. 1b-left.

Core Model. The input SOMs (i.e. 1D lattice networks with N neurons) encode timeseries samples in a distributed activity pattern, as shown in Fig. 1a. This activity pattern is generated such that the closest preferred value of a neuron to the input sample will be strongly activated and will decay, proportional with distance, for neighbouring units. The SOM specialises to represent a certain (preferred) value in the timeseries and learns its sensitivity, by updating its tuning curves shape. Given an input sample $s^p(k)$ from one timeseries at time step k, the network computes for each i-th neuron in the p-th input SOM (with preferred value $w_{in,i}^p$ and tuning curve size $\xi_i^p(k)$) the elicited neural activation as

$$a_i^p(k) = \frac{1}{\sqrt{2\pi}\xi_i^p(k)} e^{\frac{-(s^p(k) - w_{in,i}^p(k))^2}{2\xi_i^p(k)^2}}. \tag{1}$$

The winning neuron of the p-th population, $b^p(k)$, is the one which elicits the highest activation given the timeseries sample at time k

$$b^p(k) = \operatorname*{argmax}_i a_i^p(k). \tag{2}$$

The competition for highest activation in the SOM is followed by cooperation in representing the input space. Hence, given the winning neuron, $b^p(k)$, the cooperation kernel,

$$h_{b,i}^p(k) = e^{\frac{-||r_i - r_b||^2}{2\sigma(k)^2}}. \tag{3}$$

allows neighbouring neurons (i.e. found at position r_i in the network) to precisely represent the input sample given their location in the neighbourhood $\sigma(k)$ of the winning neuron. The neighbourhood width $\sigma(k)$ decays in time, to avoid twisting effects in the SOM. The cooperation kernel in Eq. 3, ensures that specific neurons in the network specialise on different areas in the input space, such that the input weights (i.e. preferred values) of the neurons are pulled closer to the input sample,

$$\Delta w_{in,i}^p(k) = \alpha(k) h_{b,i}^p(k)(s^p(k) - w_{in,i}^p(k)). \tag{4}$$

Neurons in the two SOMs are then linked by a fully (all-to-all) connected matrix of synaptic connections, where the weights in the matrix are computed using Hebbian learning. The connections between uncorrelated (or weakly correlated) neurons in each population (i.e. w_{cross}) are suppressed (i.e. darker color) while correlated neurons connections are enhanced (i.e. brighter color), as depicted in Fig. 1b-right. Formally, the connection weight $w_{cross,i,j}^p$ between neurons i, j in the different input SOMs are updated with a Hebbian learning rule as follows:

$$\Delta w_{cross,i,j}^p(k) = \eta(k)(a_i^p(k) - \bar{a}_i^p(k))(a_j^q(k) - \bar{a}_j^q(k)), \tag{5}$$

where $\bar{a}_i^p(k)$ is a "momentum" like exponential moving average. Hebbian learning ensures that when neurons fire synchronously their connection strengths increase, whereas if their firing patterns are anti-correlated the weights decrease. The weight matrix encodes the co-activation patterns between the input layers (i.e. SOMs), as shown in Fig. 1a, and, eventually, the learned growth law (i.e. functional relation) given the timeseries, as shown in Figure 1b-right.

Self-organisation and Hebbian correlation learning processes evolve simultaneously, such that both the representation and the extracted relation are continuously refined, as new samples are presented. This can be observed in the encoding and decoding functions where the input activations are projected though w_{in} (Eq. 1) to the Hebbian matrix and then decoded through w_{cross}.

Parametrization and Read-Out. In all of our experiments the data is fed to the CHIMERA which encodes each timeseries in the SOMs and learns the underlying relation in the Hebbian matrix. The SOM neural networks are responsible of bringing the timeseries in the same latent representation space where they can interact (i.e. through their internal correlation). In our experiments, each of the SOM has $N = 100$ neurons, the Hebbian connection matrix has size $N \times N$ and parametrization is done as: $\alpha = [0.01, 0.1]$ decaying, $\eta = 0.9$, $\sigma = \frac{N}{2}$ decaying following an inverse time law. We use as decoding mechanism an optimisation method that recovers the real-world value given the self-calculated bounds of the input timeseries. The bounds are obtained as minimum and maximum of a cost function of the distance between the current preferred value of the winning neuron (i.e. the value in the input which is closest to the weight vector of the neuron in Euclidian distance) and the input sample.

2.2   Datasets

In our experiments we used publicly available tumor growth datasets (see Table 2), with real clinical tumor volume measurements, for different cell lines of breast cancer. This choice is to probe and demonstrate the versatility of CHIMERA in learning from tumor growth patterns induced by different types of cancer. For the pharmacokinetics of the chemotoxic drug (i.e. Paclitaxel), we used the data from [8] describing intracellular and extracellular concentrations of Paclitaxel during uptake. MCF7 breast cancer cells were incubated with 1 to 1000 nM Paclitaxel. The concentration of Paclitaxel in cells and culture medium were monitored for 24 h (see Table 1 and Table 2 in [8]). The volume of MCF7 cells in the exponential growth phase was determined using microscopic imaging.

Table 2. Description of the breast cancer datasets used in the experiments.

Experimental dataset setup					
Dataset	Cancer type	Data type	Data points	Data freq.	Source
1	MDA-MB-231 cell line	Fluorescence imaging	7	2x/week	[13]
2	MDA-MB-435 cell line	Digital Caliper	14	2x/week	[18]
3	MCF-7 cell line	Caliper	8	1x/week	[16]
4	LM2-4LUC+ cell line	Digital Caliper	10	3x/week	[9]

2.3   Procedures

In order to reproduce the experiments in our study, the MATLAB ® code and copies of all the datasets are available on GITLAB[1]. Each of the three mechanistic tumor growth models (i.e. Logistic, Bertalanffy, Gompertz) and CHIMERA were presented the tumor growth data in each of the four datasets. When a dataset contained multiple trials and patients, a random one was chosen for testing and validation.

---

[1] https://gitlab.com/akii-microlab/chimera.

Mechanistic Models Setup. Each of the three ODE tumor growth models and the drug pharmacokinetic model was implemented as ordinary differential equation (ODE) and integrated over the dataset length. We used a solver based on a modified Rosenbrock formula of order 2 that evaluates the Jacobian during each step of the integration. To provide initial values and the best parameters (i.e. $\alpha, \beta, \lambda, k$) for each of the four models the Nelder-Mead simplex direct search (i.e. derivative-free minimum of unconstrained multi-variable functions) was used, with a termination tolerance of $10e^{-4}$ and upper bounded to 1000 iterations. Finally, fitting was performed by minimizing the sum of squared residuals.

CHIMERA Setup. For CHIMERA the data was normalized before training and de-normalized for the evaluation. The system was comprised of two input SOMs, each with $N = 50$ neurons, encoding the volume data and the irregular sampling time sequence, respectively. Both input density learning and cross-modal learning cycles were bound to 100 epochs. The full parametrization details of CHIMERA are given in Parametrization and read-out section.

3   Results

In the current section, we present the experimental results of our study and demonstrate that CHIMERA is capable to: a) learn the tumor growth model $f(V)$ from tumor growth data of breast cancer, b) learn the pharmacokinetics $P(t, V)$ of chemotoxic drug dose and c) to use the learnt quantities to provide a data-driven sequencing scheme for chemotherapy and surgery in a personalized breast cancer intervention.

3.1   Learning the Tumor Growth Function $f(V)$

The first experiment addressed the capability to learn the tumor growth model $f(V)$ from data without imposing biological constraints upon the cell line, tumor size, number of cells etc. We demonstrate the superior learning capabilities of CHIMERA on the four publicly-available breast cancer clinical datasets described in Table 2. As we can observe in Fig. 2, CHIMERA learns a superior fit (i.e. lowest Sum Squared Error (SSE), Root Mean Squared Error (RMSE) and symmetric Mean Absolute Percentage Error (sMAPE)) to the tumor growth data, despite the limited number of samples (i.e. 7 data points for MDA-MD-231 cell line dataset and up to 14 data points for MDA-MD-435 cell line dataset). Despite their ubiquitous use, the classical tumor growth models (e.g. Gompertz, von Bertalanffy, Logistic) are confined due to: a) the requirement of a precise biological description - as one can see in the different sigmoid shapes in Fig. 2; b) incapacity to describe the diversity of tumor types and; c) the small amount and irregular sampling of the data - visible in the relatively poor fit to the data, captured in Fig. 2.

3.2   Learning the Pharmacokinetics $P(t, V)$

The second experiment focused on extracting the pharmacokinetics of Paclitaxel. Its pharmacokinetics function $P(t, V)$ is dependent of its concentration $c(t)$ at time t and V the volume of the tumor. In our experiments, the drug concentration $c(t)$ has two components as following the model in [8], namely the intracellular and extracellular Paclitaxel.

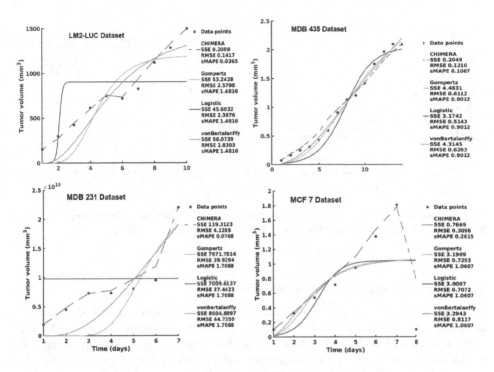

Fig. 2. Evaluation of the tumor growth models on the different datasets: accuracy evaluation. CHIMERA is decoded from the learnt Hebbian weight matrix. The decrease in the MCF7 Dataset is due to the administered chemotherapy and demonstrates the adaptivity of CHIMERA in capturing growth behaviours.

Cellular Concentration. As one can see in Fig. 3 - left panel, the intracellular concentration kinetics of Paclitaxel is highly nonlinear. CHIMERA is able to extract the underlying function describing the data without any assumption about the data and other prior information, opposite to the model from [8]. Interestingly, CHIMERA captured a relevant effect consistent with multiple Paclitaxel studies [15]. Namely, that the intracellular concentration increased with time and approached plateau levels, with the longest time to reach plateau levels at the lowest extracellular concentration - as shown in Fig. 3.

Extracellular Concentration. Analysing the extracellular concentration in Fig. 3 - right panel, we can see that CHIMERA extracted the trend and the individual variation of drug concentration after the administration of the drug (i.e. in the first 6 h) and learnt an accurate fit without any priors or other biological assumptions. Interestingly, CHIMERA captured the fact that the intracellular drug concentration increased linearly with extracellular concentration decrease, as shown in Fig. 3.

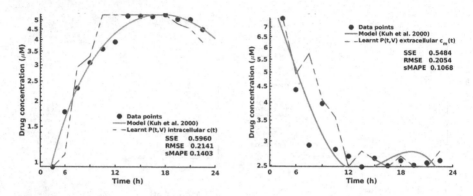

Fig. 3. Learning the pharmacokinetics $P(t, V)$ of the intracellular (left) and extracellular (right) Paclitaxel concentration $c(t)$. Data from [8], log scale plot.

3.3   Chemotherapy-Surgery Sequencing

In this sub-section we combine all the results in the previous sub-sections on learning the tumor growth function $f(V)$ and the pharmacokinetics $P(t, V)$ with a mechanistic modelling framework that demonstrates the capabilities of CHIMERA in sequencing chemotherapy and surgery in breast cancer. We evaluate the potential sequencing under the two hypotheses, log-kill and Norton-Simon, respectively. In order to simplify the formulation, we consider the tumor volume $V = Nv$, where N is the number of cells and v is a constant describing cell volume and the volume of intercellular space. We assume, for simplicity, that the growth model follows a Gompertz growth curve with a transport constant $K = e^{\frac{\beta}{\alpha}}$, $f(V) = \beta \frac{V}{v} ln(\frac{V}{vK})$ and the pharmacokinetics function is $P(t, V) = c(t)\frac{V}{v}$. Following our derivation in Sect. 1, $V_{neoadj} = e^{-k_s} V_2(t_f)$ and $V_{adj} = V_1(t_f)$ correspond to tumor sizes in neo-adjuvant and adjuvant sequences, respectively. Under the log-kill assumption, if we let $c(t) = - \int_{t_0}^{t_f} c(s)e^{\beta s} ds$ then

$$\frac{V_{neoadj}}{V_{adj}} = exp\{-k_s(1 - e^{-\beta(t_f - t_0)})\} < 1 \tag{6}$$

hence $V_{neoadj} < V_{adj}$. Similarly, under the Norton-Simon assumption we obtain

$$\frac{V_{neoadj}}{V_{adj}} = exp\{-k_s(1 - e^{-\beta(t_f - t_0)+c(t_f)})\}, \tag{7}$$

which for $c(t) = \int_{t_0}^{t_f} c(s)ds < t_f - t_0$ determines $V_{neoadj} < V_{adj}$. In analogue fashion, one can derive the sequencing for the Logistic and von Bertalanffy models by replacing the growth law with the corresponding equations in Table1. In order to evaluate the sequencing capabilities of CHIMERA with respect to traditional biologically parametrized models (we only present Gompertz in Table 3, analogue results for Logistic, von Bertalanffy), we consider the dataset of breast cancer (MCF-7 cell line) from [16] described in our Experimental setup. We use the derivations for V_{adj} and V_{neoadj} and fill in with the decoded values from the learnt tumor growth (Fig. 2) and learnt pharmacokinetics (Fig. 3). Without extensive parametrization and biological dependency, CHIMERA uses learnt tumor growth and pharmacokinetics to infer the most appropriate sequence of therapy, consistent with its mechanistic counterparts, as shown in Table 3.

Table 3. Evaluation of the chemotherapy-surgery sequencing prediction.

Model (biological parameters)	Log-kill hypothesis	Norton-Simon hypothesis
Gompertz (β, K, v)	$V_{neoadj} < V_{adj}$	$V_{neoadj} < V_{adj}{}^{a}$
CHIMERA (none)	$V_{neoadj} < V_{adj}$	$V_{neoadj} > V_{adj}$

[a] Holds only if $c(t) = \int_{t_0}^{t_f} c(s)ds < t_f - t_0$.

4 Conclusion

As ever, there will be no "one-size-fits-all" treatment for breast cancer and the focus should be on optimising patient characterization. This can be achieved through a data-driven approach in which individual patient data describing tumor growth (e.g. histology, imaging) and chemotoxic drug effect (i.e. pharmacokinetics, drug interactions) are used in combination to extract the optimal sequence of therapy. CHIMERA is an initial effort to offer a personalized solution in chemotherapy-surgery sequencing that can handle biological variability of tumors, the limited size of patient data, and the variability in chemotoxic drug response, in a data-driven way. We demonstrated and evaluated CHIMERA's capabilities in a series of experiments that emphasize the need for combining data-driven and mechanistic modelling in oncology.

References

1. Afghahi, A., et al.: Tumor BRCA1 reversion mutation arising during neoadjuvant platinum-based chemotherapy in triple-negative breast cancer is associated with therapy resistance. Clin. Cancer Res. **23**(13), 3365–3370 (2017)
2. Chen, Z., Haykin, S., Eggermont, J.J., Becker, S.: Correlative Learning: A Basis for Brain and Adaptive Systems. Wiley, Hoboken (2008)

3. Comen, E., Gilewski, T.A., Norton, L.: Tumor growth kinetics. Holland-Frei Cancer Medicine, pp. 1–11 (2016)
4. Gardner, S.N.: A mechanistic, predictive model of dose-response curves for cell cycle phase-specific and-nonspecific drugs. Cancer Res. **60**(5), 1417–1425 (2000)
5. Gerlee, P.: The model muddle: in search of tumor growth laws. Cancer Res. **73**(8), 2407–2411 (2013)
6. Kerr, D.J., Haller, D., Verweij, J.: Principles of chemotherapy. Oxford Textbook of Cancer Biology, p. 413 (2019)
7. Kohonen, T.: Self-organized formation of topologically correct feature maps. Biol. Cybern. **43**(1), 59–69 (1982)
8. Kuh, H.J., et al.: Computational model of intracellular pharmacokinetics of paclitaxel. J. Pharmacol. Exp. Ther. **293**(3), 761–770 (2000)
9. Mastri, M., Tracz, A., Ebos, J.M.: Tumor growth kinetics of human LM2-4LUC+ triple negative breast carcinoma cells (December 2019). https://doi.org/10.5281/zenodo.3574531
10. Mauri, D., Pavlidis, N., Ioannidis, J.P.: Neoadjuvant versus adjuvant systemic treatment in breast cancer: a meta-analysis. J. Natl. Cancer Inst. **97**(3), 188–194 (2005)
11. Pusztai, L., Foldi, J., Dhawan, A., DiGiovanna, M.P., Mamounas, E.P.: Changing frameworks in treatment sequencing of triple-negative and HER2-positive, early-stage breast cancers. Lancet Oncol. **20**(7), e390–e396 (2019)
12. Reid-Lawrence, S., Tan, A.R., Mayer, I.A.: Optimizing adjuvant and neoadjuvant chemotherapy for triple-negative breast cancer. In: Tan, A.R. (ed.) Triple-Negative Breast Cancer, pp. 83–94. Springer, Cham (2018). https://doi.org/10.1007/978-3-319-69980-6_7
13. Rodallec, A., Giacometti, S., Ciccolini, J., Fanciullino, R.: Tumor growth kinetics of human MDA-MB-231 cells transfected with dTomato lentivirus (December 2019). https://doi.org/10.5281/zenodo.3593919
14. Simon, R., Norton, L.: The Norton-Simon hypothesis: designing more effective and less toxic chemotherapeutic regimens. Nat. Clin. Pract. Oncol. **3**(8), 406–407 (2006)
15. Stage, T.B., Bergmann, T.K., Kroetz, D.L.: Clinical pharmacokinetics of paclitaxel monotherapy: an updated literature review. Clin. Pharmacokinet. **57**(1), 7–19 (2018)
16. Tan, G.E.A.: Combination therapy of oncolytic herpes simplex virus HF10 and bevacizumab against experimental model of human breast carcinoma xenograft. Int. J. Cancer **136**(7), 1718–1730 (2015)
17. Verhulst, P.F.: Notice sur la loi que la population suit dans son accroissement. Corresp. Math. Phys. **10**, 113–126 (1838)
18. Volk, L.D., Flister, M.J., Chihade, D., Desai, N., Trieu, V., Ran, S.: Synergy of nab-paclitaxel and bevacizumab in eradicating large orthotopic breast tumors and preexisting metastases. Neoplasia **13**(4), 327 (2011). IN14
19. Von Bertalanffy, L.: Quantitative laws in metabolism and growth. Q. Rev. Biol. **32**(3), 217–231 (1957)
20. West, J., Newton, P.K.: Chemotherapeutic dose scheduling based on tumor growth rates provides a case for low-dose metronomic high-entropy therapies. Cancer Res. **77**(23), 6717–6728 (2017)
21. de Wiel, V., et al.: Neoadjuvant systemic therapy in breast cancer: challenges and uncertainties. Eur. J. Obstet. Gynecol. Reprod. Biol. **210**, 144–156 (2017)
22. Zaheed, M., Wilcken, N., Willson, M.L., O'Connell, D.L., Goodwin, A.: Sequencing of anthracyclines and taxanes in neoadjuvant and adjuvant therapy for early breast cancer. Cochrane Database Syst. Rev. **2**(2), CD012873 (2019)

Fine-Tuning Deep Learning Architectures
for Early Detection of Oral Cancer

Roshan Alex Welikala[1]([⊠]), Paolo Remagnino[1], Jian Han Lim[2], Chee Seng Chan[2],
Senthilmani Rajendran[3], Thomas George Kallarakkal[2], Rosnah Binti Zain[2,4],
Ruwan Duminda Jayasinghe[5], Jyotsna Rimal[6], Alexander Ross Kerr[7], Rahmi Amtha[8],
Karthikeya Patil[9], Wanninayake Mudiyanselage Tilakaratne[2,5], John Gibson[10],
Sok Ching Cheong[2,3], and Sarah Ann Barman[1]

[1] Kingston University, Surrey KT1 2EE, UK
r.welikala@kingston.ac.uk
[2] University of Malaya, 50603 Kuala Lumpur, Malaysia
[3] Cancer Research Malaysia, 47500 Subang Jaya, Malaysia
[4] MAHSA University, Bandar Saujana Putra, 42610 Jenjarom, Malaysia
[5] University of Peradeniya, Peradeniya 20400, Sri Lanka
[6] BP Koirala Institute of Health Sciences, Dharan 56700, Nepal
[7] New York University, New York, NY 10010, USA
[8] Trisakti University, Kota Jakarta Barat, Jakarta 11440, Indonesia
[9] Jagadguru Sri Shivarathreeshwara University, Mysuru 570 015, Karnataka, India
[10] University of Aberdeen, Aberdeen AB25 2ZD, UK

Abstract. Oral cancer is most prevalent in low- and middle-income countries where it is associated with late diagnosis. A significant factor for this is the limited access to specialist diagnosis. The use of artificial intelligence for decision making on oral cavity images has the potential to improve cancer management and survival rates. This study forms part of the MeMoSA® (Mobile Mouth Screening Anywhere) project. In this paper, we extended on our previous deep learning work and focused on the binary image classification of 'referral' vs. 'non-referral'. Transfer learning was applied, with several common pre-trained deep convolutional neural network architectures compared for the task of fine-tuning to a small oral image dataset. Improvements to our previous work were made, with an accuracy of 80.88% achieved and a corresponding sensitivity of 85.71% and specificity of 76.42%.

Keywords: Deep learning · Oral cancer · Oral potentially malignant disorders

1 Introduction

Oral cancer is one of the most common cancers worldwide, with an estimated 354,864 new cases and 177,384 deaths in 2018 [1]. The disease disproportionately affects low- and middle-income countries (LMICs). Oral cancer is typically associated with late diagnosis, particularly in LMICs, and as a result survival rates are low [2]. Significant

© Springer Nature Switzerland AG 2020
G. Bebis et al. (Eds.): ISMCO 2020, LNBI 12508, pp. 25–31, 2020.
https://doi.org/10.1007/978-3-030-64511-3_3

factors associated with late diagnosis are poor awareness and the limited access to specialist diagnosis.

A major advantage is that oral cancer is often preceded by visible oral lesions termed as oral potentially malignant disorders (OPMDs) which can be detected from a clinical oral examination performed by a trained healthcare practitioner. Screening programs, if in place, offer early diagnosis and can lead to a reduction in mortality rates and morbidity. Telemedicine using images captured via mobile phones [3] would allow for remote consultation by specialists and may improve the referral accuracy of screening programs.

Artificial intelligence (AI) has the potential to classify images according to specific disease types or even provide descriptive summaries. However, achieving a high-level of performance for the binary classification of 'referral' vs. 'non-referral' would be the first step towards translation into clinical practice (following robust clinical evaluation). With a telemedicine approach, this would assist primary healthcare providers who may not be trained in identifying high-risk oral lesions in sending through only relevant cases to the specialists.

Recent methods related to the automated early detection of oral cancer made use of the convolutional neural network (CNN) which is a deep learning based AI technique designed for inputs in the form of images. Deep learning enables features to be automatically learnt at multiple levels of abstraction which allow complex patterns to be derived. Uthoff [4] used a CNN to classify pairs of autofluorescence and white light images as suspicious and not suspicious. Aubreville [5] used a CNN to classify laser endomicroscopy images as clinically normal and carcinogenic. Whilst custom CNN architectures can be built for a specific task, there are several popular architectures well known for achieving state-of-the-art performance on the ImageNet dataset [6] at their time of release. Among these are VGG [7], InceptionV3 [8], ResNet [9] and Xception [10].

Transfer learning is a technique where a model trained on one task is repurposed on a second related task. The biggest benefit of transfer learning shows when the target dataset is small, this is due to very large datasets being required to train deep learning models. It is common to use CNN architectures pre-trained on the ImageNet dataset which contains 1.2 million images with 1000 classes (e.g. tiger, pizza, speedboat). If a dataset is very small (e.g. <1000 images) then best practice is to use a pre-trained CNN as a fixed feature extractor, if not as small (e.g. >1000 images) then fine-tuning the CNN can produce superior results. Due to overfitting concerns with small datasets, it is advisable to keep the initial layers frozen (which capture universal low-level features such as edges, curves and blobs) and only fine-tune the latter part of the network.

Our previous work [11] focused on using ResNet to tackle early detection of oral cancer. ResNet was used to explore image classification and object detection, along with classifying according to different levels of disease categorization. In this paper, we provide a short extension to this work, focused on the binary image classification of 'referral' vs. 'non-referral'. We compared the performance of some common pre-trained CNN architectures (VGG, InceptionV3 and ResNet) when applied to our oral image dataset, whilst exploring issues of fine-tuning with respect to a small dataset.

2 Materials

This study forms part of the MeMoSA® (**M**obile **M**outh **S**creening **A**nywhere) project [3], in which images are currently in the process of being gathered and annotated from clinical experts from across the world. At this initial phase of the project, the number of annotated oral cavity images stands at 2155.

From this dataset, 1180 images were of class 'non-referral' and 975 images were of class 'referral'. The 'non-referral' class comprised of a mixture of images without lesions and images with lesions but not requiring referral. The 'referral' class comprised of images with lesions that required referral for low risk OPMD, high risk OPMD, cancer and other reasons. The images were of varying size, the largest was 5472×3648 pixels and the smallest was 119×142 pixels. The dataset was split into training, validation and test sets as detailed in Table 1. Further details on the dataset can be found in [11].

Table 1. Image numbers according to the class label and dataset type.

Class	Training	Validation	Test	Total
Non-referral	949	125	106	1180
Referral	795	82	98	975

3 Method

Five different CNN architectures were trained on our dataset for the binary image classification of 'referral' vs. 'non-referral'. The softmax layer with a 1000 outputs was changed to two outputs to represent the two classes (equivalent to sigmoid function). The training involved freezing the initial part of the networks and fine-tuning the latter part of the networks, which included the convolutional layers responsible for high-level features. These architectures are detailed in Table 2; the stated top1/top5 accuracies were reported by Keras [12] for performance of the ImageNet dataset.

Table 2. CNN architectures.

Architecture	Description	Top-1	Top-5
VGG-16	13 convolutional and 3 fully-connected (FC) layers (including the softmax layer). Its novelty was to go deeper	71.3%	90.1%
VGG-19	A deeper variant to VGG-16	71.3%	90.0%
Inception-V3	48 layers with no FC layers except for the softmax layer. Its novelty was the concatenation of feature maps generated by filters of multiple sizes. Among the first to use batch normalization	77.9%	93.7%
ResNet-50	50 layers with no FC layers except for the softmax layer. Its novelty was to popularize skip connections with residual blocks to combat training issues associated with very deep networks. Among the first to use batch normalization	74.9%	92.1%
ResNet-101	A deeper variant to ResNet-50	76.4%	92.8%

3.1 Technical Details

Backpropagation and stochastic gradient descent (SGD) with momentum of 0.9 was used for training. Images were rescaled to 224 × 224 pixels, except for InceptionV3 which used 299 × 299 pixels. The training data was augmented with horizontal/vertical flipping, scaling, translation and rotation.

SGD mini-batch size was 128 images. A weighted loss function was used to correct for the slight class imbalance in the training data. The models were initialized with pre-trained ImageNet weights and fine-tuned from the second last convolutional layer for VGG, from second last Inception block for InceptionV3 and from conv4_1 for ResNet. The training strategy varied based on the architecture, e.g. VGG-19 was trained for 100 epochs at a learning rate of 0.001. Varying levels of weight decay were used for regularization. The models were built on the training set and hyperparameters were derived from performance on the validation set.

A Nvidia GeForce RTX 2080 Ti graphics card with 11 GB memory was used for training. This implementation used Keras and TensorFlow.

3.2 Batch Normalization for Transfer Learning

Batch Normalization (BN) targets the vanishing gradient problem by standardizing the output of the previous layer, it speeds up the training process and it enables deeper networks to be trained. During training BN uses the mean and variance of the current mini-batch to normalize, and during inference BN uses fixed batch statistics derived from the moving mean and variance that was estimated during training.

BN works well when fine-tuning the entire network. But when part of the network is frozen (due to limited data) the behavior of BN can cause discrepancies between training and inference. Consider the frozen part of the network; for training BN uses the current mini-batch statistics and for inference BN uses fixed batch statistics derived from the original dataset. This works well if the data is from the same/similar domain as ImageNet, but leads to poor results if the domain is different (i.e. oral cancer). This was rectified when BN in the frozen part of the network was set to use moving mean and variance that was estimated during training for both training and inference.

An additional issue, when the dataset is small and the domain is different, is to achieve representative fixed batch statistics (used for inference) for the data. Training for long enough resolves this, but this is problematic with limited data. We find a BN momentum value of 0.9 helped towards achieving better statistics.

These issues affected the IncpetionV3 and ResNet models which used BN throughout their architecture.

4 Results

Evaluation was performed on the test set. As the classes were approximately balanced in the test set, we used accuracy as a single performance metric to compare the architectures (as detailed in Table 3). For each architecture, a confidence score threshold that produced the best operating point defined by the accuracy was selected. The best performing architecture was VGG19 with an accuracy of 80.88%. This corresponds to a sensitivity of 85.71% and a specificity of 76.42%, with further metrics detailed in Table 4. Examples of outputs from VGG-19 are provided in Fig. 1.

Table 3. Image classification results for several CNN architectures.

Architecture	Accuracy (%)	Architecture	Accuracy (%)
VGG-16	80.39	ResNet-50	74.51
VGG-19	80.88	ResNet-101	76.96
Inception-V3	76.47		

Table 4. Further metrics on the image classification results for VGG-19.

Performance metric	(%)	Performance metric	(%)
Sensitivity	85.71	False positive rate	23.58
Specificity	76.42	Precision	77.06
Positive predictive value	77.06	Recall	85.71
Negative predictive value	85.26	F_1 score	81.16
False negative rate	14.29	Accuracy	80.88

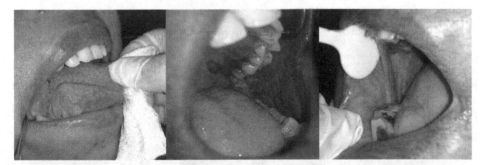

Fig. 1. Examples of results. Left: correctly classified as 'non-referral' with a class probability of 0.82. Middle: correctly classified as 'referral' with a class probability of 0.85. Right: incorrectly classified as 'non-referral' with a class probability of 0.83.

5   Discussion and Conclusion

In this paper, we demonstrate the performance of deep learning based systems for the image classification of 'referral' vs. 'non-referral' with respect to oral cancer. The best performing model achieved a sensitivity of 85.71% and a specificity of 76.42% for the identification of images that required referral. An accuracy of 80.88% and F_1 score of 81.16%; it surpassed the F_1 score of 78.30% reported in previous work [11].

After exploring several CNN architectures, we demonstrate that the VGG architectures produced superior results for our dataset, with the VGG-19 coming out on top.

Despite InceptionV3 and ResNet being more complex and deeper networks, they were surpassed by the easier to fine-tune VGG models. Aside from overfitting issues with deeper networks (although VGG does contain FC layers which can also cause overfitting); InceptionV3 and ResNet present batch normalization issues when being used for fine-tuning on small datasets of a different domain. Therefore, the VGG models currently provide a more stable and reliable approach, better showing the potential of AI. However, the other architectures offer more potential to learn complex patterns and will be used to produce superior results when our dataset is larger.

Our future scope is to pre-train our models on datasets from a similar domain as our data (e.g. skin cancer), this is likely to improve results. But the most important target is to build a large dataset as this is key to deep learning. This will enable fine-tuning of the entire network, or even training the architectures from scratch and training custom made architectures. We plan to focus on the interpretability of our models to support clinical confidence in AI decision making, briefly covered in the appendix.

In conclusion, we have shown potential for AI to be incorporated into a mobile phone based telemedicine approach for the early detection of oral cancer. These promising early results are set to improve as the MeMoSA® project continues and the dataset grows.

Acknowledgments. We would like to thank the Medical Research Council for providing funding (MR/S013865/1).

Appendix

Fig. 2. Correctly classified as 'referral' with a class probability of 0.787. Left: original image. Right: Grad-CAM for class 'referral'.

Figures 2 and 3 provide gradient-weighted class activation mappings (Grad-CAM) [13] to demonstrate where the VGG-19 model was looking. The model appears to approximately focus on the lesion when making the decision of 'referral'. We feel our models could benefit from using a trainable attention mechanism [14].

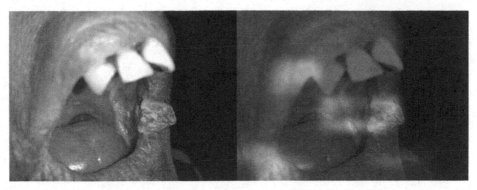

Fig. 3. Correctly classified as 'referral' with a class probability of 0.791. Left: original image. Right: Grad-CAM for class 'referral'.

References

1. Bray, F., Ferlay, J., Soerjomataram, I., Siegel, R.L., Torre, L.A., Jemal, A.: Global cancer statistics 2018: GLOBOCAN estimates of incidence and mortality worldwide for 36 cancers in 185 countries. CA: Cancer J. Clin. **68**(6), 394–424 (2018)
2. Doss, J.G., Thomson, W.M., Drummond, B.K., Latifah, R.J.R.: Validity of the FACT-H&N (v 4.0) among Malaysian oral cancer patients. Oral Oncol. **47**(7), 648–652 (2011)
3. Haron, N., et al.: m-Health for early detection of oral cancer in low-and middle-income countries. Telemed. e-Health **26**(3), 278–285 (2020)
4. Uthoff, R.D., et al.: Point-of-care, smartphone-based, dual-modality, dual-view, oral cancer screening device with neural network classification for low-resource communities. PLoS One **13**(12), e0207493 (2018)
5. Aubreville, M., et al.: Automatic classification of cancerous tissue in laserendomicroscopy images of the oral cavity using deep learning. Sci. Rep. **7**(1), 1–10 (2017)
6. Deng, J., Dong, W., Socher, R., Li, L.-J., Li, K., Fei-Fei, L.: ImageNet: a large-scale hierarchical image database. In: IEEE Conference on Computer Vision and Pattern Recognition, pp. 248–255 (2009)
7. Simonyan, K., Zisserman, A.: Very deep convolutional networks for large-scale image recognition. arXiv Preprint arXiv:1409.1556 (2014)
8. Szegedy, C., Vanhoucke, V., Ioffe, S., Shlens, J., Wojna, Z.: Rethinking the inception architecture for computer vision. In: IEEE Conference on Computer Vision and Pattern Recognition, pp. 2818–2826 (2016)
9. He, K., Zhang, X., Ren, S., Sun, J.: Deep residual learning for image recognition. In: IEEE Conference on Computer Vision and Pattern Recognition, pp. 770–778 (2016)
10. Chollet, F.: Xception: deep learning with depthwise separable convolutions. In: IEEE Conference on Computer Vision and Pattern Recognition, pp. 1251–1258 (2017)
11. Welikala, R., et al.: Automated detection and classification of oral lesions using deep learning for early detection of oral cancer. IEEE Access **8**, 132677–132693 (2020)
12. Keras. https://keras.io/api/applications/. Accessed 22 June 2020
13. Selvaraju, R.R., Cogswell, M., Das, A., Vedantam, R., Parikh, D., Batra, D.: Grad-CAM: visual explanations from deep networks via gradient-based localization. Int. J. Comput. Vis. **128**(2), 336–359 (2019). https://doi.org/10.1007/s11263-019-01228-7
14. Jetley, S., Lord, N.A., Lee, N., Torr, P.H.S.: Learn to pay attention. arXiv preprint arXiv:1804.02391 (2018)

Discriminative Localized Sparse Representations for Breast Cancer Screening

Sokratis Makrogiannis[(⊠)] [iD], Chelsea E. Harris[iD], and Keni Zheng[iD]

Delaware State University, Dover, DE 19901, USA
smakrogiannis@desu.edu

Abstract. Breast cancer is the most common cancer among women both in developed and developing countries. Early detection and diagnosis of breast cancer may reduce its mortality and improve the quality of life. Computer-aided detection (CADx) and computer-aided diagnosis (CAD) techniques have shown promise for reducing the burden of human expert reading and improve the accuracy and reproducibility of results. Sparse analysis techniques have produced relevant results for representing and recognizing imaging patterns. In this work we propose a method for Label Consistent Spatially Localized Ensemble Sparse Analysis (LC-SLESA). In this work we apply dictionary learning to our block based sparse analysis method to classify breast lesions as benign or malignant. The performance of our method in conjunction with LC-KSVD dictionary learning is evaluated using 10-, 20-, and 30-fold cross validation on the MIAS dataset. Our results indicate that the proposed sparse analyses may be a useful component for breast cancer screening applications.

1 Introduction

In this paper we introduce a method for classification of breast masses into benign and malignant states. Breast cancer diagnosis is a significant area of research [6], and because of its significance, automated detection and diagnosis of breast cancer is a popular field of research [2,13,15,17,26]. Early diagnosis has been shown to reduce mortality and significantly improve quality of life. To achieve this, mammograms are used to help detect breast cancer at an early stage. Detection and diagnosis of breast cancer requires high levels of expertise and experience, and is carried out by trained radiologists. Computer-aided diagnosis would reduce the time for diagnosis and improve its reproducibility.

A widely adopted method for diagnosis and early prediction of breast cancer is the X-ray mammographic test [12]. Hence the research field of computer-aided detection and diagnosis for breast cancer has attracted significant interest. Conventional classification models such as those introduced in [3,7,14,18,20,23] use specific rules to craft features. Texture, shape and intensity features were extracted in [20]. Among the extracted features, genetic algorithm (GA) selected the most appropriate features. Zernike moments have also been used to extract features due to their ability to well describe the shape of objects [21]. In more recent years, state of the art techniques using neural networks have been used to extract and select features [11]. Among these

G. Bebis et al. (Eds.): ISMCO 2020, LNBI 12508, pp. 32–43, 2020.
https://doi.org/10.1007/978-3-030-64511-3_4

techniques, Convolutional Neural Nets (CNNs) are a popular choice. Key advances in design and application of CNNs [10,24] significantly improved the state of the art in object recognition for the imagenet dataset. A commonly used training strategy for neural networks for medical imaging classification applications is transfer learning [8,11,30]. In [5], for example, pretrained VGG16, ResNet50, and Inception v3 networks were customized to be applied on different datasets.

This research focuses on the diagnosis (CADx) of breast cancer lesions as benign or malignant using sparse representation and dictionary learning. Sparse representation has found wide applicability in signal/image processing, computer vision and pattern recognition. Sparse representation methods seek to represent signals using sparse linear approximations of patterns, or atoms that compose the dictionary matrix. These sparse approximations can be used for compression and denoising of signals/images, classification, object recognition, and other applications. A central theme in such techniques is dictionary learning, which studies methods for learning dictionaries that lead to optimal representations according to the application objective. These methods have produced impressive results in various signal and image processing tasks [1,9,19,25,28,29,31,32]. In the recent years, convolutional sparse coding and its relationship with deep learning methods has been widely studied [4,19,29].

Despite the interest in these techniques, their application to biomedical field is still somewhat limited to straightforward application of sparse representation classification, or learning of multiple separate dictionaries, one for each class followed by representation residual-based loss functions. Therefore there is still motivation for the design of methods that leverage the capabilities of dictionary learning and sparse coding using joint discriminative-generative approaches.

In this work we introduce discriminative dictionary techniques that utilize class label consistency into spatially localized ensemble sparse analysis classification that we denote as LC-SLESA. We utilize this method for breast cancer diagnosis in mammograms. Our premise is that spatially localized dictionaries that have been optimized using label consistency constraints [9], will improve the classification accuracy of spatially localized sparse analysis.

1.1 Sparse Analysis

Conventional sparse representation of signals has been a research area of considerable interest in recent years. In image classification tasks, the representation of an image is used to determine the class of the test image. Sparse analysis seeks to optimize an signal representation objective function with signal sparsity constraints. This function contains a term that measures the difference between the image reconstruction and the test image. This difference is known as the reconstruction error or residual. A sparsity term is also included in the objective function to measure the sparsity of the computed solution. The reconstruction error term may be set to measure the test image exactly or a constraint may be defined to provide an upper bound for the reconstruction error. The success of sparse representation based classification is due to the fact that a high-dimensional image can be represented or coded by few representative samples. Sparse representation based classification has two phases: coding and classification. Within the coding phase an image/signal is collaboratively coded over a dictionary of atom with some sparsity constraint. In the classification performed based on the coding coefficients and the dictionary.

The dictionary could be predefined. For example, a dictionary may be made up of all training samples from all classes. This type of dictionary may not be able to represent test samples well due to the uncertain and noisy information in the original training images. Large number of dictionary atoms increase the coding complexity.

The optimization problem that sparse analysis seeks to solve is as follows: Given samples from space $\mathbb{R}^d$, a dictionary $D \in \mathbb{R}^{d \times n}$ of signals partitioned by class, and a test signal $y \in \mathbb{R}^d$, sparse coding seeks to find a coding vector $x \in \mathbb{R}^n$. The test signal y will be represented as a linear combination of the dictionary atoms with respect to the solution vector

$$y = Dx \tag{1}$$

This is an ill-posed inverse problem. The problem is regularized by assuming sparse solutions; the sparsity of x ultimately ensures that recovery is possible. Sparsity is represented by the l_0 norm and can be approximated by the l_1 norm, or l_p norms with $p \in (0, 1)$. Sparse coding seeks to find a solution $\hat{x}$ by solving the following problem,

$$\hat{x} = \arg\min_x ||\hat{x}||_0 \quad subject \quad to \quad y = Dx \tag{2}$$

Assuming a noisy signal, we can introduce the approximation tolerance ε and solve the following problem,

$$\hat{x} = \arg\min_x ||\hat{x}||_0 \quad subject \quad to \quad ||y - Dx|| < \varepsilon \tag{3}$$

The objective function in Eq. (2) or (3) is optimized through sparse coding. Conventional sparse representation methods such as SRC [27], optimize an objective function of two terms, and use the original training images as the atoms in D. More recent works emphasize on dictionary design and task-specific optimization that we discuss next.

1.2 Dictionary Learning

Sparse representation seeks to represent a signal by linear combinations of dictionary atoms. While the objective is to achieve sparse solutions, the dictionary is a key factor that drives the optimization problem. The design of dictionary from training data has gained significant interest in recent years [22,28]. This area, known as dictionary learning seeks to construct the best dictionary possible to provide more efficient representation of classes for a specific task. Learning the dictionary has shown to lead to better representations versus using an unlearned dictionary [28]. Dictionary learning can be categorized into three major areas [25]: (i) probabilistic learning methods, (ii) clustering-based learning methods, and (iii) construction methods. The signal that will be represented, denoted by y, is of size $1 \times d$, the solution $\hat{x}$ is of size $n \times 1$ where n is the number of atoms within the dictionary, and the dictionary D is of size $d \times n$.

The dictionary (sometimes referred to as the 'resource' database) consists of signals, also called atoms. Each atom makes up a column of the dictionary. The type, size, and design of the dictionary are important features. A dictionary may be pre-selected or trained (learned). The size of the dictionary has shown to have much effect on the representation formed. If the number of atoms within the dictionary (n) exceeds the dimension of the test signal (d), that is if $d < n$, the dictionary is said to be overcomplete. Studies show that an overcomplete/redundant dictionary leads to more sparse representation of signals [16].

2 Label Consistent Spatially Localized Ensemble Sparse Analysis (LC-SLESA)

Based on our previous work on spatially localized ensemble sparse analysis (SLESA), here we introduce label consistency into the localized dictionaries and we denote this method by LC-SLESA. Our SLESA approach reduces the dimension of the feature vector, and integrate Bayesian decision learners to correct the estimation bias. At the same time, the ensemble classification we designed builds an integrated model by calculating the sparse representation of a block structure, thereby determining the lesion category. LC-SLESA aims to further improve the performance of SLESA by finding task-specific dictionaries that are consistent with the class labels of the training data.

2.1 Spatially Localized Block Decomposition

We divide each image into blocks with size $m \times n$, represented as $I = [B^1, B^2, ..., B^{NBL}]$, where B^j denotes a block of each training image, and NBL is the number blocks of an image. The dictionary D^j, where $j = 1, 2, ..., NBL$, is corresponding to the same position of the block B^j for all s images:

$$D^j = [B_1^j, B_2^j, \cdots, B_s^j]. \tag{4}$$

2.2 Block-Based Label Consistent KSVD for Dictionary Learning

After the decomposition into spatially localized blocks, we learn NBL discriminative dictionaries using the label consistent KSVD algorithm (denoted by LC-KSVD) proposed in [9]. This method adds task-specific constraints to the sparse representation problem to compute a single discriminative dictionary.

In this work, we employ the LC-KSVD method to learn spatially localized dictionaries D^j. We employed two variations of the objective function; LC-KSVD1 and LC-KSVD2. LC-KSVD1 adds a label consistent regularization term for the objective function and solves the following problem

$$\arg \min_{D^j, A^j, x^j} ||Y^j - D^j X^j||_2^2 + ||Q^j - A^j X^j||_2^2 \quad s.t. \quad ||x_m^j||_0 \leq T, \quad \forall m \in [1, n]. \tag{5}$$

The added term is the discriminative sparse code error that forces signals from the same class to have similar sparse codes. Q^j denotes the discriminative sparse codes for Y^j, A^j is a linear transformation matrix, and T is the sparsity threshold.

Furthermore, LC-KSVD2 adds a joint classification error and label consistent regularization term to the objective function. LC-KSVD2 is defined as

$$\arg \min_{D^j, A^j, W^j, x^j} ||Y^j - D^j X^j||_2^2 + ||Q^j - A^j X^j||_2^2 + ||H^j - W^j X^j||_2^2 \quad s.t. \quad ||x_m^j||_0 \leq T. \tag{6}$$

The new term represents the classification error, W^j denotes linear classifier parameters, and H^j contains the class labels of the training data Y^j [9].

2.3 Ensemble Classification

At this stage we combine the individual spatially localized decisions to classify the test samples. For each test sample y^j in jth block, we find the solution x^j of the regularized noisy l_1-minimization problem:

$$\widehat{x}^j = \arg\min \|x^j\|_1 \text{ subject to } \|D^j x - y^j\|_2 \leq \varepsilon \tag{7}$$

We propose ensemble learning techniques in a Bayesian probabilistic setting as weighted sums of classifier predictions. We propose a decision function that applies majority voting to individual hypotheses (BBMAP) and an ensemble of log-likelihood scores computed from relative sparsity scores (BBLL).

Maximum a Posteriori Decision Function (BBMAP). The class label of each test image determined by the overall vote of the NBL blocks-based classifiers. The predicted class label $\widehat{\omega}$ is given by

$$\widehat{\omega}_{BBMAP} = \mathscr{F}_{BBMAP}(\widehat{x}) \doteq \arg\max_i pr(\omega_i|\widehat{x}), \tag{8}$$

The probability for classifying $\widehat{x}$ into class ω_i is

$$pr(\omega_i|\widehat{x}) = \sum_j^{NB} ND_{\omega_i^j}/NB \tag{9}$$

$$ND_{\omega_i^j} = \begin{cases} 1, & \text{if } \widehat{x}^j \in i \text{ th class} \\ 0, & \text{otherwise} \end{cases}, \tag{10}$$

where $ND_{\omega_i^j}$ is an indicator function, and its value is determined by the decision of each classifier.

Log Likelihood Sparsity-Based Decision Function (BBLL-S). A likelihood score based on the relative sparsity scores $\|\delta_m(\widehat{x}^j)\|_1$, $\|\delta_n(\widehat{x}^j)\|_1$ calculated at the sparse representation stage of each classifier

$$LLS(\widehat{x}) = -\log \frac{\|\delta_m(\widehat{x}^j)\|_1}{\|\delta_n(\widehat{x}^j)\|_1} \begin{cases} \geq 0, & \widehat{x}^j \in m\text{th class} \\ < 0, & \widehat{x}^j \in n\text{th class} \end{cases}. \tag{11}$$

The expectation of $LLS(\widehat{x})$ for all classifiers that we denote by $ELLS$ is estimated over the individual classification scores obtained by (11)

$$ELLS \doteq E\{LLS(\widehat{x})\} = \frac{1}{NB} \sum_j^{NB} LLS(\widehat{x}^j)$$

$$= -\frac{1}{NB} \left[\sum_j^{NB} \log \|\delta_m(\widehat{x}^j)\|_1 - \sum_j^{NB} \log \|\delta_n(\widehat{x}^j)\|_1 \right], \tag{12}$$

We apply a sigmoid function $\varsigma(.)$ to determine the state of $\hat{\omega}$ through the decision threshold τ_{LLS}.

$$\hat{\omega}_{BBLL} = \mathscr{F}_{BBLL}(\hat{x}) \doteq \varsigma(ELLS(\hat{x}) - \tau_{LLS}). \tag{13}$$

3 Experiments and Discussion

In this section we describe our experiments and report results produced by our approach and by widely used convolutional neural networks [10,24] to accommodate comparisons.

3.1 Data

We evaluated our CAD techniques for separation of breast lesions into two classes: malignant and benign. The training and testing data were obtained from the Mammographic Image Analysis Society (MIAS) database that is available online [15]. The resolution of the mammograms is 200 micron pixel edge that corresponds to about 264.58 μm pixel size, and the size of each image is 1024×1024 px after clipping/padding. MIAS contains 322 MLO scans from 161 subjects. Our goal is to characterize the lesion type, therefore we utilized 66 benign and 51 malignant mammograms for performance evaluation.

ROI selection is applied first, in order to prepare the data for block decomposition. We need to ensure that the majority of the blocks cover the lesion to improve the accuracy. Hence, we designed our system so that the lesion ROI sizes are greater than or equal to the analysis ROI size. Our method reads-in the centroid and radius of each mass from the provided radiological readings. It uses these two values to automatically determine a minimum bounding square ROI and to select the masses that satisfy the size criterion. We used a ROI size criterion of 64×64 px, resulting in 36 benign and 37 malignant lesions. These ROIs contain sufficient visual information, while preserving a big part of the data samples. We performed 10-, 20- and 30-fold cross-validation to study the effect of the size of folds on performance.

3.2 Convolutional Neural Networks

In this part of our experiments we implemented CNN classification using the Alexnet [10] and Googlenet [24] architectures with transfer learning. Both networks were pretrained on Imagenet that is a database of 1.2 million natural images.

We applied transfer learning to each network in slightly different ways. To adjust Alexnet to our data, we replaced the pre-trained fully connected layers with three new fully connected layers. We set the learning rates of the pre-trained layers to 0 to keep the network weights fixed, and we trained the new fully connected layers only. In the case of Googlenet, we set the learning rates of the bottom 10 layers to 0, we replaced the top fully connected layer with a new fully connected layer, and we assigned a greater learning rate factor for the new layer than the pre-trained layers.

To provide the networks with additional training examples, we employed data resampling using randomly-centered patches, followed by data augmentation by rotation, scaling, horizontal flipping, and vertical flipping. Finally, we applied hyperparameter tuning using Bayesian optimization to find the optimal learning rate, mini-batch size and number of epochs.

We first applied 10-, 20-, and 30-fold cross-validation to 64×64 px ROIs. Because deep networks can learn information from the edges of lesions and not just the texture, we also decided to test our method on 256×256 px ROIs of all lesions (66 benign and 51 malignant) to improve the classification performance. We report the results of our cross-validation experiments in Table 1. We note that Alexnet produces the top ACC of 67.65% and the top AUC of 63.04% for 30-fold cross-validation and for all 256×256 px ROIs.

Table 1. Classification performance for breast lesion characterization using convolutional neural network classifiers (ROI size: 64×64)

Method	k-Fold CV	ROI Size	TPR (%)	TNR (%)	ACC (%)	AUC (%)
Alexnet	10	64×64	50.0	58.33	54.17	47.69
		256×256	56.86	72.55	**64.71**	**62.19**
	20	64×64	44.44	69.44	56.94	52.55
		256×256	47.06	84.31	**65.69**	**60.7**
	30	64×64	38.89	72.22	55.56	53.97
		256×256	58.82	64.71	61.77	60.29
Googlenet	10	64×64	25.0	83.33	54.17	47.21
		256×256	64.71	58.82	61.77	57.86
	20	64×64	58.33	50.0	54.17	50.22
		256×256	62.75	62.75	62.75	61.5
	30	64×64	58.33	50.00	54.17	50.96
		256×256	66.67	68.63	**67.65**	**63.04**

3.3 LC-SLESA

Next, we validated our block-based ensemble classification system. We performed 10-, 20- and 30-fold cross-validation on these samples. We present results on characterization of lesions with minimum ROI size of 64×64 pixels that forms a dataset of 36 benign and 37 malignant lesions.

Table 2 contains the classification rates produced by our cross-validation experiments for multiple block sizes. In the first row of this table, the results were obtained from a single block that is equivalent to conventional SRC analysis [27]. We note that the accuracy increases when the number of folds increases for the same ROI size. The highest accuracy by using 10-fold cross-validation is 75.71% for 32×32 block size

Table 2. Classification performance for breast lesion characterization using ensembles of block-based sparse classifiers with dictionary learning (ROI size: 64×64, random state permutation)

Method	k-Fold CV	Block Size	SLESA ACC %	SLESA AUC %	LC-SLESA1 ACC (%)	LC-SLESA1 AUC (%)	LC-SLESA2 ACC (%)	LC-SLESA2 AUC (%)
BBMAP-S	10	64×64	57.14	54.55	47.14	53.51	51.43	47.50
		32×32	64.29	64.29	67.14	68.14	55.71	54.63
		16×16	70.00	69.70	70.00	69.70	70.00	67.70
		8×8	64.29	63.64	70.00	69.70	70.00	69.70
BBLL-S	10	64×64	60.00	62.33	55.71	42.51	57.14	52.74
		32×32	64.29	60.44	**75.71**	73.96	62.86	57.66
		16×16	68.57	70.84	72.86	73.96	70.00	73.96
		8×8	70.00	71.42	71.40	73.87	70.00	**77.31**
BBMAP-S	20	64×64	46.67	42.38	45.00	40.16	60.00	55.73
		32×32	58.33	54.84	53.33	50.17	55.00	52.84
		16×16	75.00	76.64	76.67	76.42	71.67	71.08
		8×8	75.00	76.64	70.00	69.19	75.00	75.19
BBLL-S	20	64×64	63.33	53.73	53.33	48.05	61.67	58.18
		32×32	60.00	55.95	55.00	54.62	68.33	61.07
		16×16	76.67	78.87	**80.00**	**88.88**	76.67	80.65
		8×8	76.66	81.20	78.33	86.99	78.33	86.21
BBMAP-S	30	64×64	51.67	46.50	46.67	40.71	56.67	50.61
		32×32	53.33	48.83	60.00	55.51	50.00	46.38
		16×16	85.00	85.65	85.00	85.65	85.00	85.65
		8×8	83.33	82.09	70.00	69.19	85.00	85.65
BBLL-S	30	64×64	65.00	58.62	53.33	43.83	60.00	61.18
		32×32	68.33	65.41	66.67	60.07	66.67	62.63
		16×16	86.67	88.21	**88.33**	93.66	83.33	88.88
		8×8	83.33	89.10	85.00	94.66	85.00	**95.88**

for LC-SLESA1 with BBLL decision function. The largest area under the curve for 10-fold CV is 77.31% for 8×8 block size for LC-SLESA2 with BBLL decision function. For 20-fold cross-validation, the best accuracy is 80.00% and AUC is 88.88% for 8×8 block size for LC-SLESA1 with BBLL decision function. The best overall performance is obtained for 30-fold cross validation. The highest accuracy is 88.33% for 16×16 block size for LC-SLESA1 with BBLL decision function, and the largest area under the curve is 95.88% for 8×8 block size for LC-SLESA2 with BBLL decision function. There are 2 or 3 test samples in each fold when $k = 30$. We also display the receiver operating curves by variations of the SLESA method in Fig. 1 for 8×8 and 16×16 block lengths. These graphs lead to the same observations that we made from Table 2. In addition, Fig. 2 displays an example of dictionaries based on the original training images, learned by LS-KSVD1 and learned by LC-KSVD2. We see that the label consistent algorithms learn structural features of the lesions.

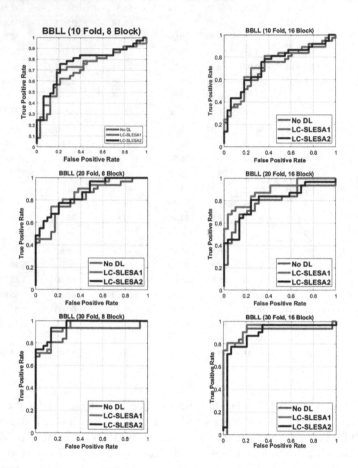

Fig. 1. ROC plots for 8×8 and 16×16 block sizes using the proposed block-based ensemble method with BBLL decision functions and 10-fold (top row), 20-fold (second row) and 30-fold (bottom row) cross-validation.

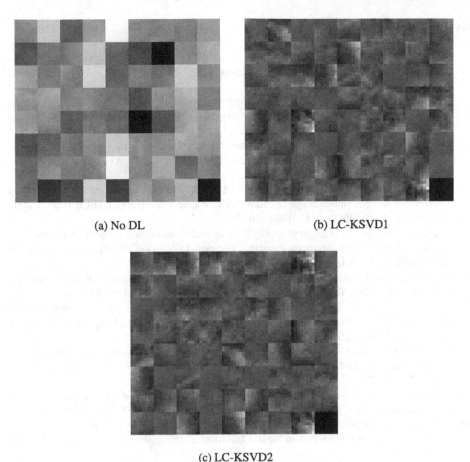

(a) No DL   (b) LC-KSVD1

(c) LC-KSVD2

Fig. 2. Dictionary comparison example for (a) SLESA without dictionary learning, (b) LC-SLESA1, and (c) LC-SLESA2 methods.

4   Conclusion

We proposed discriminative localized sparse representations for classifying breast masses into benign and malignant states. This approach introduces discriminative capability into the generative method of sparse representation for classification. As we observed in our experiments, this approach improves the classification accuracy of integrative sparse analysis and accomplishes an area under the ROC of about 95.88% for 30-fold cross-validation.

Acknowledgments. The authors acknowledge the support by the National Institute of General Medical Sciences of the National Institutes of Health under Award Number SC3GM113754. They also acknowledge the support by Delaware CTR-ACCEL (NIH U54GM104941) and the State of Delaware.

References

1. Aharon, M., Elad, M., Bruckstein, A.: K-svd: an algorithm for designing overcomplete dictionaries for sparse representation. IEEE Trans. Signal Process. **54**(11), 4311–4322 (2006). https://doi.org/10.1109/TSP.2006.881199
2. Benjamin, Q.H., Hui, L.M.L.G.: Digital mammographic tumor classification using transfer learning from deep convolutional neural networks. J. Med. Imag. **3**(3), 1–5 (2016). https://doi.org/10.1117/1.JMI.3.3.034501
3. Beura, S., Majhi, B., Dash, R.: Mammogram classification using two dimensional discrete wavelet transform and gray-level co-occurrence matrix for detection of breast cancer. Neurocomputing **154**, 1–14 (2015)
4. Chang, H., Han, J., Zhong, C., Snijders, A.M., Mao, J.H.: Unsupervised transfer learning via multi-scale convolutional sparse coding for biomedical applications. IEEE Trans. Pattern Anal. Mach. Intell. **40**(5), 1182–1194 (2017)
5. Chougrad, H., Zouaki, H., Alheyane, O.: Deep convolutional neural networks for breast cancer screening. Comput. Methods Programs Biomed. **157**, 19–30 (2018). https://doi.org/10.1016/j.cmpb.2018.01.011
6. Ferlay, J., Héry, C., Autier, P., Sankaranarayanan, R.: Global Burden of Breast Cancer, pp. 1–19. Springer, New York (2010). https://doi.org/10.1007/978-1-4419-0685-41
7. George, M., Chen, Z., Zwiggelaar, R.: Multiscale connected chain topological modelling for microcalcification classification. Comput. Biol. Med. **114**, 103422 (2019). https://doi.org/10.1016/j.compbiomed.2019.103422
8. Hepsağ, P.U., Özel, S.A., Yazıcı, A.: Using deep learning for mammography classification. In: 2017 International Conference on Computer Science and Engineering (UBMK), pp. 418–423. IEEE (2017)
9. Jiang, Z., Lin, Z., Davis, L.S.: Label consistent k-svd: learning a discriminative dictionary for recognition. IEEE Trans. Pattern Anal. Mach. Intell. **35**(11), 2651–2664 (2013)
10. Krizhevsky, A., Sutskever, I., Hinton, G.E.: Imagenet classification with deep convolutional neural networks. In: Advances in Neural Information Processing Systems, pp. 1097–1105 (2012)
11. Litjens, G., et al.: A survey on deep learning in medical image analysis. Med. Image Anal. **42**, 60–88 (2017)
12. Misra, S., Solomon, N.L., Moffat, F.L., Koniaris, L.G.: Screening criteria for breast cancer. Adv. Surgery **44**(1), 87–100 (2010)
13. Nagarajan, R., Upreti, M.: An ensemble predictive modeling framework for breast cancer classification. Methods **131**, 128–134 (2017). https://doi.org/10.1016/j.ymeth.2017.07.011, systems Approaches for Identifying Disease Genes and Drug Targets
14. Narváez, F., Alvarez, J., Garcia-Arteaga, J.D., Tarquino, J., Romero, E.: Characterizing architectural distortion in mammograms by linear saliency. J. Med. Syst. **41**(2), 1–12 (2016). https://doi.org/10.1007/s10916-016-0672-5
15. Oliver, A., et al.: A review of automatic mass detection and segmentation in mammographic images. Med. Image Anal. **14**(2), 87–110 (2010)
16. Papyan, V., Romano, Y., Elad, M.: Convolutional neural networks analyzed via convolutional sparse coding. J. Mach. Learn. Res. **18**(1), 2887–2938 (2017)
17. Pereira, D.C., Ramos, R.P., Do Nascimento, M.Z.: Segmentation and detection of breast cancer in mammograms combining wavelet analysis and genetic algorithm. Comput. Methods and Programs Biomed. **114**(1), 88–101 (2014)
18. Rabidas, R., Midya, A., Chakraborty, J.: Neighborhood structural similarity mapping for the classification of masses in mammograms. IEEE J. Biomed. Health Inform. **22**(3), 826–834 (2017)

19. Rey-Otero, I., Sulam, J., Elad, M.: Variations on the convolutional sparse coding model. IEEE Trans. Signal Process. **68**, 519–528 (2020)

20. Rouhi, R., Jafari, M., Kasaei, S., Keshavarzian, P.: Benign and malignant breast tumors classification based on region growing and cnn segmentation. Expert Syst. Appl. **42**(3), 990–1002 (2015)

21. Sharma, M.K., Jas, M., Karale, V., Sadhu, A., Mukhopadhyay, S.: Mammogram segmentation using multi-atlas deformable registration. Comput. Biol. Med. **110**, 244–253 (2019). https://doi.org/10.1016/j.compbiomed.2019.06.001

22. Shrivastava, A., Patel, V.M., Pillai, J.K., Chellappa, R.: Generalized dictionaries for multiple instance learning. Int. J. Comput. Vis. **114**(2–3), 288–305 (2015)

23. Singh, S.P., Urooj, S.: An improved CAD system for breast cancer diagnosis based on generalized pseudo-zernike moment and Ada-DEWNN classifier. J. Med. Syst. **40**(4), 1–13 (2016). https://doi.org/10.1007/s10916-016-0454-0

24. Szegedy, C., et al.: Going deeper with convolutions. In: Computer Vision and Pattern Recognition (CVPR) (2015), http://arxiv.org/abs/1409.4842

25. Tosic, I., Frossard, P.: Dictionary learning. IEEE Signal Process. Magazine **28**(2), 27–38 (2011)

26. Verma, B., McLeod, P., Klevansky, A.: Classification of benign and malignant patterns in digital mammograms for the diagnosis of breast cancer. Expert Syst. Appl. **37**(4), 3344–3351 (2010)

27. Wright, J., Yang, A.Y., Ganesh, A., Sastry, S.S., Ma, Y.: Robust face recognition via sparse representation. IEEE Trans. Pattern Anal. Mach. Intell. **31**(2), 210–227 (2009). https://doi.org/10.1109/TPAMI.2008.79

28. Yang, M., Zhang, L., Feng, X., Zhang, D.: Sparse representation based fisher discrimination dictionary learning for image classification. Int. J. Comput. Vis. **109**(3), 209–232 (2014)

29. Zhang, Z., Xu, Y., Yang, J., Li, X., Zhang, D.: A survey of sparse representation: algorithms and applications. IEEE Access **3**, 490–530 (2015)

30. Zhao, W., Xu, R., Hirano, Y., Tachibana, R., Kido, S.: A sparse representation based method to classify pulmonary patterns of diffuse lung diseases. Comput. Math. Methods Med. **2015** (2015). https://doi.org/10.1155/2015/567932

31. Zheng, K., Makrogiannis, S.: Sparse representation using block decomposition for characterization of imaging patterns. In: Wu, G., Munsell, B.C., Zhan, Y., Bai, W., Sanroma, G., Coupé, P. (eds.) Patch-Based Techniques in Medical Imaging: Third International Workshop, Patch-MI 2017, Held in Conjunction with MICCAI 2017, Quebec City, QC, Canada, September 14, 2017, Proceedings, pp. 158–166. Springer, Cham (2017). https://doi.org/10.1007/978-3-319-67434-6_18

32. Zhou, Y., Chang, H., Barner, K., Spellman, P., Parvin, B.: Classification of histology sections via multispectral convolutional sparse coding. In: Proceedings of the IEEE Conference on Computer Vision and Pattern Recognition, pp. 3081–3088 (2014)

Activation vs. Organization: Prognostic Implications of T and B Cell Features of the PDAC Microenvironment

Elliot Gray[1] , Shannon Liudahl[2], Shamilene Sivagnanam[1,2], Courtney Betts[2],
Jason Link[4,6], Dove Keith[4], Brett Sheppard[4,7], Rosalie Sears[4,5,6],
Guillaume Thibault[3], Joe W. Gray[3,5], Lisa M. Coussens[2,4,5],
and Young Hwan Chang[1,3(✉)]

[1] Computational Biology and Department of Biomedical Engineering,
Oregon Health and Science University (OHSU), Portland, OR, USA
{grael,chanyo}@ohsu.edu
[2] Department of Cell, Developmental, and Cancer Biology, OHSU,
Portland, OR, USA
[3] OHSU Center for Spatial Systems Biomedicine, Portland, OR, USA
[4] Brenden-Colson Center for Pancreatic Care, Portland, OR, USA
[5] Knight Cancer Institute, Portland, OR, USA
[6] Department of Molecular and Medical Genetics, OHSU, Portland, OR, USA
[7] Department of Surgery, OHSU, Portland, OR, USA

Abstract. Pancreatic ductal adenocarcinoma (PDAC) patients, who
often present with stage III or IV disease, face a dismal prognosis as the 5-
year survival rate remains below 10%. Recent studies have revealed that
$CD4^+$ T, $CD8^+$ T, and/or B cells in specific spatial arrangements relative
to intratumoral regions correlate with clinical outcome for patients, but
the complex functional states of those immune cell types remain to be
incorporated into prognostic biomarker studies. Here, we developed an
interpretable machine learning model to analyze the functional relationship between leukocyte-leukocyte or leukocyte-tumor cell spatial proximity, correlated with clinical outcome of 46 therapy-naïve PDAC patients
following surgical resection. Using a multiplex immunohistochemistry
imaging data set focused on profiling leukocyte functional status, our
model identified features that distinguished patients in the fourth quartile from those in the first quartile of survival. The top ranked important
features identified by our model, all of which were positive prognostic
stratifiers, included CD4 T helper cell frequency among $CD45^+$ immune
cells, frequency of Granzyme B-positivity among CD4 and CD8 T cells,
as well as the frequency of PD-1 positivity among CD8 T cells. The spatial proximity of CD4 T- to B cells, and between CD8 T cells and epithelial cells, were also identified as important prognostic features. While spatial proximity features provided valuable prognostic information, the best
model required both spatial and phenotypic information about tumor
infiltrating leukocytes. Our analysis links the immune microenvironment
of PDAC tumors to outcome of patients, thus identifying features associated with more progressive disease.

© Springer Nature Switzerland AG 2020
G. Bebis et al. (Eds.): ISMCO 2020, LNBI 12508, pp. 44–55, 2020.
https://doi.org/10.1007/978-3-030-64511-3_5

Keywords: Pancreatic cancer · Multiplexed imaging · Machine learning

1 Introduction

Pancreatic ductal adenocarcinoma (PDAC), the most common form of pancreatic cancer, is predicted to become the second leading cause of cancer death in the US by 2030 [17]. While the survival rate for patients with stage I disease is showing an improving trend, the vast majority of patients continue to present with stage II or later disease and a 5-year expected survival probability less than 20% (even lower for stages III or IV) [2]. Worse, few patients present with targetable mutations (7% with BRCA1/BRCA2) [17], and checkpoint blockade immunotherapies have not shown much efficacy [20]. Surgical resection and adjuvant chemotherapy offers survival benefit typically measured in months, and 80% of patients will go on to relapse and die from their disease [17]. To help these patients, new treatments improving overall survival are needed.

PDAC is characterized histologically by the presence of a dense, fibrotic, inflamed stroma that modulates patient response to chemotherapy [20]. As part of this, an adaptive immune response marked by the accumulation of $CD8^+$ T cells is often present, and tends to confer an improved prognosis [6,10,19]. Importantly, analysis of patients from the neoadjuvant GVAX trial [18] revealed that vaccine treatment could stimulate the formation of a robust adaptive immune response, but that myeloid cells inhibited T cell activation and reduced overall survival [24]. Also, the spatial organization of T cells, B cells, and epithelial cells in the tumor microenvironment has been previously correlated with survival [5–7,16,19,19], reflecting the cell signaling events that occur at various distance scales (i.e. direct touching to diffusion of signaling molecules over longer distances).

Building on these studies, we asked whether similar T cell and myeloid signatures could be found in PDAC patients receiving standard of care surgery plus adjuvant chemotherapy, to confirm whether the same immunosuppressive mechanisms were at play without vaccine treatment. Furthermore, given the aforementioned findings relating spatial localization of T cells to patient prognosis, we wanted to place our analysis of T cell activation or suppression in the context of their spatial microenvironment, to understand the relative importance of these factors with regards to patient survival. To do this, we performed a study of 46 human PDAC tissues using multiplexed immunohistochemistry [24] (mIHC).

Following quantification of single-cell data from the mIHC images, we assessed the relative prognostic value of features using a machine learning model. We compared the spatial proximity of leukocyte subsets, to each other and to the tumor epithelium, and the functional state of those leukocytes as defined by markers of proliferation, cytotoxicity, and activation. While a few previous studies have applied multiplexed imaging and spatial pattern analysis to the study of PDAC [6,19], ours makes use of an expanded antibody panel including a rich categorization of leukocytes. Our study supports the utility of mIHC-based immuno-profiling as a biomarker discovery tool, and sheds light on the interplay between adaptive immune response and immunosuppresion in PDAC patients receiving standard of care treatment.

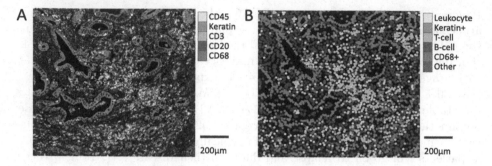

Fig. 1. A) Representative mIHC image showing a density of leukocytes within pathologist-annotated tumor. **B)** Scatter plot showing cellular coordinates (centroid of nucleus), colored by identified cell type, resulting from quantification of the image shown in part A. The cell type definitions used in this study are given in Table 1. (Color figure online)

2   Methods

2.1   Multiplexed Immunohistochemistry

Following the protocol defined previously [24], we subjected archival 4–5 μm formalin-fixed parrafin-embedded PDAC tissue specimens to sequential staining and imaging, using antibodies that enabled precise identification of epithelial cells and functionally distinct subsets of T and B cells (see Table 1). We performed image registration, single-cell segmentation, and quantification of pixelwise-mean intensity of chromogen per cell as previously described [24], the results of which can be seen in Fig. 1 A & B. Lastly, we used manual gating, following the classification hierarchy laid out in Table 1, to classify cells into discrete types.

With the help of an expert pathologist, we identified the borders of the tumor, and selected image regions of interest (ROI) within the boundaries of tumor annotation. We specifically did not include any region that was identified as a tertiary lymphoid structure (TLS) [15] within the ROIs, but selected the ROIs to maximize the sampling of leukocyte dense regions that did not suffer tissue loss during staining. Patients had between 1–4 ROIs, and the combined areas of those ROIs were $22.6\,mm^2$ on average ($22.6\,mm^2$ median) with 90% of patients having between 5 and $36\,mm^2$ of quantified tumor area.

For each of the quantified cell types or functional states in Table 1, we defined their frequency among their parent leukocyte subset (Table 1, last column), for

Table 1. Cell type classification & quantification scheme for each cell type. Note: Pan-Keratin antibody abbreviated as "Keratin".

Cell Type	Protein Marker	Frequency Denominator
Epithelium	Keratin$^+$	n/a
Leukocyte	Keratin$^-$ CD45$^+$	n/a
CD4$^+$/CD8$^+$ T cell	Leukocyte, CD3$^+$CD4$^+$/8$^+$	# Leukocytes
B cell	Leukocyte, CD3$^-$CD20$^+$	# Leukocytes
Monocyte	Leukocyte, CD68$^+$	# Leukocytes
T follicular helper	CD4T, PD1$^+$	# CD4$^+$ T cells
Function	**Protein Marker**	**Frequency Denominator**
Proliferation	Ki67	# CD4$^+$ or CD8$^+$ T cells
Cytotoxicity	Granzyme B (Gzmb)	# CD4$^+$/CD8$^+$ T cells
B cell activation	CD27$^+$/$^-$, IgD$^+$/$^-$	# B cells
CD8 T cell activation	Eomes	# CD8 T cells
Early/Late Effector	PD1	# Eomes$^+$/$^-$ CD8T
Immunoregulatory	ICOS	# PD1$^+$CD4 T cells
Variable	**Numerator**	**Denominator**
CD8$^+$ T to CD68$^+$ ratio	# CD8$^+$ T cells	# CD68$^+$ Leukocytes

example the frequency of CD4$^+$ T cells among all leukocytes or the frequency of Granzyme B-positivity among CD8$^+$ T cells. Additionally, guided by previous observations of the prognostic interplay between CD8$^+$ T cells and CD68$^+$ myeloid cells [24], we included their ratio as an additional feature in our modeling (Table 1, last row).

2.2   Patient Meta-data and Target Variables

The patient cohort we studied contained one-hundred and four patients with PDAC, who received no presurgical chemotherapy, whose tumors were surgically resected, and who went on to receive standard-of-care chemotherapy. The patients were treated at two institutions, 46 of which are included in this study (discovery cohort), and the remaining 58 of which will form a validation cohort in our future study. We performed staining of all tissue samples in the same physical research laboratory at OHSU.

We wanted to identify features that were strongly associated with survival time, so we made the decision *a priori* to compare the patients with the most extreme positive and negative outcomes. To this end, we categorized the patients into Short (1st quartile), Medium (2nd or 3rd quartile), or Long (4th quartile) survival groups, and subsequently compared Short (S) to Long (L). In the discovery cohort, there were 6 patients in the long-term survival group, and 15 in the short-term survival group. Note that in the discovery cohort, none of the patients in the short-term survival group had a censored survival time.

2.3 Quantification of Spatial Proximity Between Cell Types

Following the approach of Goltsev *et al.* [13], we quantified the degree of mixing between pairs of cell types at predefined distance scales as the log-likelihood ratio (*LLR*) of counts of pairs of neighboring cells. Briefly, for a given radius R, a graph is constructed in which each pair of cells is connected by an edge if their spatial locations are closer than R in distance. We define N as the total number of edges in the graph, N_A as the number of edges incident to nodes of type A, and N_{AB} as the number of edges connecting nodes of type A to nodes of type B. The *LLR* of proximity between cells of types A and B is then defined in Eq. (1) as:

$$LLR_{AB}(R) = log_{10}\frac{observed}{expected} = log_{10}\frac{N(R) \cdot N_{AB}(R)}{N_A(R) \cdot N_B(R)} \tag{1}$$

For comparison, we give the definition of the bivariate Ripley's K function [6,9], in the context of the same graph as defined above, as follows:

$$K_{AB}(R) = \frac{area \cdot N_{AB}(R)}{n_A \cdot n_B} \tag{2}$$

where the number of nodes of types A and B, respectively, are n_A and n_B, and the area of the image or study region is *area*.

LLR is similar in interpretation to (edge-effect corrected) Ripley's K, and indeed is correlated with it in this data (data not shown), but is different in that the denominator quantifies expected proximity between A and B conditioned on the graph structure, while Ripley's K denominator quantifies expected proximity under the condition of complete spatial randomness [9]. The *LLR* might be considered to be a more realistic model of spatial mixing, given that it incorporates information about the overall tissue architecture, however in practice the two measures are similar. We also note that both the *LLR* and Ripley's K (before edge-effect correction) are symmetric, i.e. $LLR_{AB} = LLR_{BA}$ and $K_{AB} = K_{BA}$ by definition.

2.4 Predictive Model of Survival Group

We consider a binary classification problem (i.e., long-term survivors vs short-term survivors). For our model we used Extremely Randomized Trees [12] (ExtraTrees), as implemented in Scikit-Learn [4]. This algorithm is a variant of Random Forests [3] that builds an ensemble of decision trees, each of which contributes a vote to the classification of a patient. We chose this model because it is capable of learning non-linear associations and theoretically robust to over-fitting in the presence of uninformative or redundant features [12]. In our experience using simulated data (study not shown), ExtraTrees compares favorably to Random Forests, Elastic Net [25], and Support Vector Machines[14] in terms of classification accuracy in similar problem settings.

We used 500 trees and tuned parameters controlling tree depth by maximizing the leave-one-out cross-validation sum of receiver-operating-characteristic area-under-the-curve (ROC-AUC) and average-precision (AP) scores on the training set. We tuned the model by Gaussian process optimization with the BayesOpt python package [21]. Finally, we computed test scores by leave-one-out cross validation.

3 Results

3.1 Prognostic Value of *LLR* Metric Depends on Spatial Scale

First, we explored the distribution of *LLR* proximity values across all patients, without regard for clinical outcome, to confirm that the *LLR* metric of cell type proximity could recapitulate known aspects of PDAC tumor biology. For instance, Keratin$^+$ epithelial cells adhere to each other via desmosomes to form ducts or glands, so we would expect that *LLR* values quantifying epithelial-to-epithelial proximity would be positive. Furthermore, T cells and B cells form aggregates in PDAC that are thought to function as spatial platforms for regulation of T cell activation [15], so we would expect that in at least some of the tumors the *LLR* values for T cell or B cell proximity to each other would also be positive. Indeed, in Fig. 2A it is evident that these trends hold true in general, validating the ability of the *LLR* metric (with $R = 40\,\mu\text{m}$) to quantify the presence of functional epithelial and leukocyte structures.

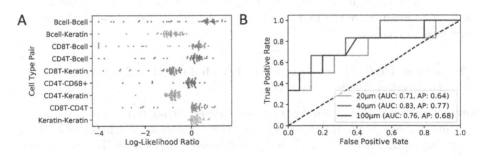

Fig. 2. A) Spatial proximity between each pair of cell types, measured by *LLR*, is shown with a dot for each patient (n = 46). *LLR* was measured with $R = 40\ \mu\text{m}$. **B)** Performance of Extra Trees model on three feature sets, defined in Table 2 caption as *"Func-LLR"*, each of which was produced with a different value of R used in computing the *LLR* variables.

According the the *LLR* metric, the proximity of CD8$^+$ T cells to Pan-Keratin$^+$ tumor cells shown in Fig. 2A is negative for all the patients. This is in agreement with prior observations that stromal CD8$^+$ T cell density is higher than that in the tumor core [19]. Given the typical median survival time of the patients in our study (median 20.4 months), relative to the population of

PDAC patients in the USA, we interpret that if $CD8^+$ T cell proximity to tumor is really prognostic, then moderately negative LLR values should correspond to longer survival.

When training our model to predict patient survival, we restricted the quantification of proximity between cell types to those pairs which we hypothesized to influence prognosis. Cell type pairs we used for survival analysis are listed in Table 2 in the *Base-LLR* panel. We thought that B cell proximity to T cells or other B cells, or conversely to Keratin$^+$ epithelium (indicating B cell dispersion), would reflect either a positive or negative prognosis, respectively. This might be due to the role of B cells as professional antigen presenting cells, and is supported by previous findings relating B cell spatial distribution to survival in PDAC [7]. We included proximity between T cells and Keratin$^+$ epithelium, given the cytotoxic potential of these lymphocytes. We also included proximity between $CD4^+$ T and $CD8^+$ T cells, since $CD4^+$ T helper cells provide co-stimulation necessary for T cell activation, and finally, we included the proximity between $CD68^+$ myeloid cells and $CD4^+$ T cells, given previous findings in studies of breast cancer mouse models that showed that Th2-polarized $CD4^+$ T cells licence macrophages to promote tumor survival and metastasis [8,23].

In addition to choosing between metrics (such as LLR or Ripley's K), when measuring proximity one must also choose a characteristic spatial distance. Scales in the 20–100 μm range [6,10] have been chosen in the past as representative of the range at which the probability of paracrine signalling requiring cell-to-cell contact would be highest. To reduce the bias inherent in our modeling, we compared the prognostic value of models trained with LLR values taken with R equal to 20, 40, and 100 μm. Shown in Fig. 2B, we found that using $R = 40$ μm gave the highest scores, so we proceeded to use that value for downstream analysis.

3.2 40 μm LLR Synergizes with T Cell Functional Markers to Predict Patient Survival

Next, we asked which of our features was important for model predictions, and whether there was any difference in the utility of spatial proximity (LLR) features versus the features quantifying T and B cell functionality. To address these questions, we evaluated our prognostic model with four groups of features representing broadly distinct biological hypotheses. As shown in Table 2, group *Base* is a baseline set of features containing simple metrics of immune composition. Group *Func* combines the baseline group *Base* with the measurements of the frequency of each T or B cell functional state among the parent cell types. Group *Base-LLR* instead combines the baseline with the LLR measurements ($R = 40$ μm) as we defined in the Sect. 3.1, while group *Func-LLR* is the union of *Base-LLR* and *Func*.

To control for the different sizes of these groups of features, we used the importance-based feature ranking from the initial fit of Extra Trees to perform recursive feature elimination (RFE). In each step of RFE, the least important feature is removed from the set, and the model is re-trained and evaluated using

Table 2. Definition of feature groups used in Fig. 3A. Group *Func-LLR* is the union of groups *Base-LLR* and *Func*. In group *Func*, parentheses indicate the denominator for frequency calculation; see also Table 1.

Base	%B cell	*Func*	*Base*, plus:
(n = 5)	%CD4 T	(n = 17)	%Ki67 (CD8T)
	%CD8 T		%Ki67 (CD4T)
	%CD68$^+$		%Gzmb (CD8T)
	CD8T/CD68$^+$		%Gzmb (CD4T)
Base-LLR	*Base*, plus:		%PD1 (CD4T)
(n = 12)	*LLR* B cell - B cell		%ICOS (PD1$^+$CD4T)
	LLR B cell - CD4T		%Eomes (CD8T)
	LLR B cell - CD8T		%PD1 (Eomes$^+$CD8T)
	LLR B cell - Keratin		%PD1 (Eomes$^-$CD8T)
	LLR CD8T - Keratin		%CD27$^+$IgD$^-$ (B cell)
	LLR CD8T - CD4T		%CD27$^+$IgD$^+$ (B cell)
	LLR CD4T - CD68$^+$		%CD27$^-$IgD$^+$ (B cell)

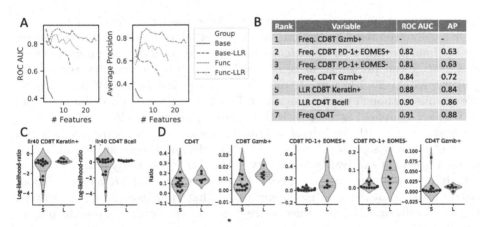

Fig. 3. A) Comparison of model performance by recursive feature elimination given different groups of features. Groups are detailed in Table 2. **B)** The optimal subset of features identified from *Func-LLR* feature group, ranked by the gain importance [3] of the Extra Trees model trained (and cross-validated) with all 24 features. ROC AUC and average precision (AP) scores are those of a model trained with all features ranked less than or equal to the variable in that row. **C)** The *LLR* values for CD8T cell to Keratin$^+$ cells, and for CD4T to B cells, were top ranked in group *Base-LLR* and also group *Func-LLR*, indicating their prognostic value. **D)** Additionally, five T cell features from groups *Func* and *Func-LLR* were top ranked in both groups. Here, the distributions of these features are shown conditioned on patient outcome, expressed as short (S) or long (L) overall survival. We chose not to perform univariate tests in C or D, and show these distributions to illustrate the positive/negative contribution of these features to patient survival in the context of our model.

cross-validation, producing ROC AUC and AP scores associated with that feature subset (Fig. 3A,B). This analysis indicates that maximal prognostic potential is achieved when both spatial LLR measurements and T cell functional markers are assessed simultaneously. Interestingly, neither the frequency of CD8$^+$ T cells, nor the ratio of CD8$^+$ T to CD68$^+$ myeloid cells were in the best feature subset, which ran contrary to our hypothesis. This result supports the hypothesis that it is the cytotoxic potential, rather then the sheer abundance, of CD8$^+$ T cells that supports patient survival.

In Fig. 3B, one can see that high ROC-AUC is achieved with only 4 features taken from the $Func$ group, but that high average precision is not achieved until additional features from the $Base$-LLR group are included. The characteristic high ROC-AUC/low AP score for the model ranked 4 in Fig. 3B represents an accurate but imprecise model. The increase in average precision from 0.72 to 0.86 by inclusion of LLR features in the model (lines 5 and 6 in Fig. 3B) indicates a substantially reduced tendency for false-positive predictions when considering both T cell functional and spatial LLR features.

3.3   Patients with Low CD4$^+$ T-to-B Cell or CD8$^+$ T-to-Keratin$^+$ Proximity Had Poor Outcomes

Using our machine learning approach, we determined that the LLR values for CD8T cell to Keratin$^+$ epithelium, and for CD4$^+$ T cell to B cell were necessary for optimal stratification of patients by survival (Fig. 3B). The survival benefit of CD8$^+$ T cell proximity to tumor, either by density within the tumor core [19], or within 20–100 μm of tumor cells [6,10], has been well documented. Meanwhile, although co-localization of T cells and B cells is a defining feature of tertiary lymphoid structures [11,15], given that we specifically filtered out those structures with the help of a trained pathologist, this finding may reflect a distinct mode of signalling between CD4$^+$ T cells and B cells in the tumor that promotes patient survival. Additionally, we note that the utility of this feature is largely driven by the presence of outlier patients with low LLR values and poor outcomes (Fig. 3C, second plot), suggesting that enhanced signalling between CD4$^+$ T cells and B cells is necessary, but not sufficient, to promote patient survival.

3.4   T Cell Cytotoxicity and CD8$^+$ T Cell Effector Frequency Are Elevated in Long-Term Survivors

Lastly, our feature importance analysis indicated that the frequency of Granzyme B positivity on both CD4 and CD8 T cells, the frequency of PD1 positivity on both Eomes$^+$ and Eomes$^-$ CD8T cells, and the overall frequency of CD4 T cells could be used to predict patient survival with high ROC AUC and AP scores (Fig. 3B), and were all elevated among the long term survivors (Fig. 3D). Although PD1$^+$Eomes$^+$ and PD1$^+$Eomes$^-$ CD8$^+$ T cells represent functionally distinct T cell phenotypes, the positive prognostic association for both features is supported by the interpretation of PD1 frequency as marking the induction of a tumor-specific immune response [22]. These findings are consistent with the theory that cytotoxicity and effector status of CD8T cells are biological markers of an efficacious anti-tumor immune response [1].

4   Discussion

As researchers and clinicians strive to bring novel therapeutic options to patients with PDAC, understanding of the biology of the immune microenvironment will be critical. In particular, identification of the cell types and proximity relationships that predict improved (or worsened) patient outcomes under standard of care provides evidence of their clinical relevance that informs future clinical trial design and drug development. Multiplex imaging technologies, which profile protein expression while retaining tissue spatial organization, allow identification of such relationships, but proof-of-concept studies taking advantage of high-dimensional imaging in PDAC still remain few in number [6,19,24]. In this study, we compared the prognostic value of measuring leukocyte spatial organization and functional activation in the context of PDAC tumors. Our finding that both spatial and functional features are required to produce the highest scoring model underscores the importance of quantifying not just what cells are doing, but also where they are doing it.

Limitations of this study are the small sample size and the use of cross-validation to evaluate predictive generalizability. We also introduced potential bias by choosing the particular spatial scales and definitions of cell state frequencies, although we did this in an effort to reduce the dimensionality of the data set *a priori*. In the future, we will validate our findings with an additional group of patients treated at an independent institution. Additionally, we plan to investigate more deeply the associations between spatial organization of cells and their functional activation states, which may give insights into the leukocyte regulatory processes that are associated with particular tissue-level organization phenotypes.

Funding and Ethical Statement. YHC acknowledges funding from the NIH (U54CA209988, 1U01 CA224012). The study and analyses were funded by an AACR Stand Up to Cancer grant funded by the Lustgarten Foundation, and the Brenden-Colson Center for Pancreatic Care at OHSU. LMC acknowledges funding from the National Institutes of Health (1U01 CA224012, U2C CA233280, R01 CA223150, R01 CA226909, R21 HD099367), the Knight Cancer Institute, and the Brenden-Colson Center for Pancreatic Care at OHSU. RCS acknowledges funding from the NIH (1U01 CA224012, U2C CA233280, U54 CA209988, R01 CA196228, and R01 CA186241) and the Brenden-Colson Center for Pancreatic Care at OHSU. This study was also made possible with support from the Oregon Clinical & Translational Research Institute (OCTRI), which is supported by the National Center for Advancing Translational Sciences, National Institutes of Health, through Grant Award Number UL1TR002369.
Human PDAC surgical resection specimens were obtained in accordance with the Declaration of Helsinki and acquired with IRB approval from Dana-Farber/Harvard Cancer Center, and the Oregon Pancreas Tissue Registry under OHSU IRB protocol # 3609.

Additional PDAC archival surgical resection specimens were collected from consented patients enrolled in the multi-center phase 1b PRINCE clinical trial (NCT03214250, sponsored by Parker Institute for Cancer Immunotherapy).

References

1. Binnewies, M., et al.: Understanding the tumor immune microenvironment (TIME) for effective therapy. Nat. Med. **24**(5), 541–550 (2018). https://doi.org/10.1038/s41591-018-0014-x
2. Blackford, A.L., Canto, M.I., Klein, A.P., Hruban, R.H., Goggins, M.: Recent trends in the incidence and survival of stage 1a pancreatic cancer: a surveillance, epidemiology, and end results analysis. JNCI J. Nat. Cancer Inst. **112**, 11629–1169 (2020)
3. Breiman, L.: Random forests. Mach. Learn. **45**(1), 5–32 (2001). https://doi.org/10.1023/A:1010933404324
4. Buitinck, L., et al.: API design for machine learning software: experiences from the scikit-learn project. arXiv preprint arXiv:1309.0238 (2013)
5. Cabrita, R., et al.: Tertiary lymphoid structures improve immunotherapy and survival in melanoma. Nature **577**, 561–565 (2020). https://doi.org/10.1038/s41586-019-1914-8
6. Carstens, J.L., et al.: Spatial computation of intratumoral T cells correlates with survival of patients with pancreatic cancer. Nat. Commun. **8**, 15095 (2017)
7. Castino, G.F., et al.: Spatial distribution of B cells predicts prognosis in human pancreatic adenocarcinoma. Oncoimmunology **5**(4), e1085147 (2016). https://doi.org/10.1080/2162402X.2015.1085147
8. Denardo, D.G., et al.: CD4+ T cells regulate pulmonary metastasis of mammary carcinomas by enhancing protumor properties of macrophages. Cancer Cell **16**(2), 91–102 (2009). https://doi.org/10.1016/j.ccr.2009.06.018
9. Dixon, P.M.: Ripley's K Function. Statistics Reference Online, Wiley StatsRef (2014)
10. Ene-Obong, A., et al.: Activated pancreatic stellate cells sequester CD8+ T cells to reduce their infiltration of the juxtatumoral compartment of pancreatic ductal adenocarcinoma. Gastroenterology **145**(5), 1121–1132 (2013)
11. Fridman, W.H., Zitvogel, L., Sautes-Fridman, C., Kroemer, G.: The immune contexture in cancer prognosis and treatment. Nat. Rev. Clin. Oncol. **14**(12), 717–734 (2017). https://doi.org/10.1038/nrclinonc.2017.101
12. Geurts, P., Ernst, D., Wehenkel, L.: Extremely randomized trees. Mach. Learn. **63**(1), 3–42 (2006). https://doi.org/10.1007/s10994-006-6226-1
13. Goltsev, Y., et al.: Deep profiling of mouse splenic architecture with CODEX multiplexed imaging. Cell **174**(4), 968–981 (2018)
14. Guyon, I., Weston, J., Barnhill, S., Vapnik, V.: Gene selection for cancer classification using support vector machines. Mach. Learn. **46**(1), 389–422 (2002). https://doi.org/10.1023/A:1012487302797
15. Hiraoka, N., Ino, Y., Yamazaki-Itoh, R., Kanai, Y., Kosuge, T., Shimada, K.: Intratumoral tertiary lymphoid organ is a favourable prognosticator in patients with pancreatic cancer. Br. J. Cancer **112**(11), 1782 (2015)
16. Jackson, H.W., Fischer, J.R., Zanotelli, V.R.T., Ali, H.R., Weber, W.P., Bodenmiller, B.: The single-cell pathology landscape of breast cancer. Nature **578**, 615–620 (2019). https://doi.org/10.1038/s41586-019-1876-x

17. Kleeff, J., et al.: Pancreatic cancer. Nat. Rev. Dis. Primers **2**, 1–23 (2016). https://doi.org/10.1038/nrdp.2016.22

18. Lutz, E.R., et al.: Immunotherapy converts nonimmunogenic pancreatic tumors into immunogenic foci of immune regulation. Cancer Immunol. Res. **2**(7), 616–631 (2014)

19. Masugi, Y., et al.: Characterization of spatial distribution of tumor-infiltrating CD8+ T cells refines their prognostic utility for pancreatic cancer survival. Mod. Pathol. (2019). https://doi.org/10.1038/s41379-019-0291-z

20. Neesse, A., et al.: Stromal biology and therapy in pancreatic cancer: ready for clinical translation? Gut **68**(1), 159–171 (2019)

21. Nogueira, F.: Bayesian optimization: open source constrained global optimization tool for Python (2014)

22. Schietinger, A., Greenberg, P.D.: Tolerance and exhaustion: defining mechanisms of T cell dysfunction. Trends Immunol. **35**(2), 51–60 (2014)

23. Shiao, S.L., Ruffell, B., DeNardo, D.G., Faddegon, B.A., Park, C.C., Coussens, L.M.: TH2-polarized CD4+ T cells and macrophages limit efficacy of radiotherapy. Cancer Immunol. Res. **3**(5), 518–525 (2015)

24. Tsujikawa, T., et al.: Quantitative multiplex immunohistochemistry reveals myeloid-inflamed tumor-immune complexity associated with poor prognosis. Cell Rep. **19**(1), 203–217 (2017). https://doi.org/10.1016/j.celrep.2017.03.037

25. Zou, H., Hastie, T.: Regularization and variable selection via the elastic net. J. Roy. Stat. Soci.: Ser. B (Stat. Methodol.) **67**(2), 301–320 (2005)

On the Use of Neural Networks
with Censored Time-to-Event Data

Elvire Roblin[1,2,3](✉) (iD), Paul-Henry Cournede[3] (iD), and Stefan Michiels[1,2] (iD)

[1] Department of Biostatistics and Epidemiology,
Gustave Roussy, University Paris-Saclay, Villejuif, France
elvire.roblin@gustaveroussy.fr
[2] Oncostat U1018, Inserm, University Paris-Saclay, labeled Ligue Contre le Cancer,
Villejuif, France
[3] MICS - Laboratory of Mathematics and Computer Science, CentraleSupélec,
University of Paris-Saclay, Gif-sur-Yvette, France

Abstract. Motivation: The objective of this work is to confront artificial neural network models with time-to-event data, using specific ways to handle censored observations such as pseudo-observations and tailored loss functions.

Methods: Different neural network models were compared. Cox-CC (Kvamme et al., 2019) uses a loss function based on a case-control approximation. DeepHit (Lee et al., 2019) is a model that estimates the probability mass function and combines log-likelihood with a ranking loss. DNNSurv (Zhao et al., 2019) circumvents the problem of censoring by using pseudo-observations. We also proposed other ways of computing pseudo-observations. We investigated the prediction ability of these models using data simulated from an Accelerated Failure Time model (Friedman et al., 2001), with different censoring rates. We simulated 100 data sets of 4,000 samples and 20 variables each, with pairwise interactions and non-linear effects of random subsets of these variables. Models were compared using the concordance index and integrated Brier score. We applied the methods to the METABRIC breast cancer data set, including 1,960 patients, 6 clinical covariates and the expression of 863 genes.

Results: In the simulation study, we obtained the highest c-indices and lower integrated Brier score with CoxTime for low censoring and pseudo-discrete with high censoring. On the METABRIC data, the neural networks obtained comparable 5-year and 10-year discrimination performances with slightly higher values for the models based on optimised pseudo-observations.

Availability: https://github.com/eroblin/NN_Pseudobs

Keywords: Survival data · Neural networks · Pseudo-observations

1 Introduction

Recently Artificial Neural Networks have been increasingly used to model complex patterns in various fields. As they are flexible non-linear models, they may

© Springer Nature Switzerland AG 2020
G. Bebis et al. (Eds.): ISMCO 2020, LNBI 12508, pp. 56–67, 2020.
https://doi.org/10.1007/978-3-030-64511-3_6

be particularly relevant when a high number of candidate covariates with non-linear effects and interactions are to be evaluated. In medical research one of the most common types of clinical outcomes is time-to-event data, such as survival times or time to clinical deterioration. Therefore, we wanted to study the ability of neural networks to learn interactions between particular patient characteristics and survival. However, we are often confronted with the problem of incomplete data, notably right-censoring, when a subject leaves the study or is lost to follow-up before the event of interest occurs or the study ends before the event has occurred. Handling censoring is a key point. To circumvent the issue of right-censored data, several approaches have recently been developed.

In this work, we compared existing survival models based on neural networks and various loss functions, and measured the robustness of these different propositions for handling survival data. As neural networks are supposed to better harvest complex patterns, we analyzed the predictive ability of the different models on a simulated data set with strong non-linear effects and interactions. Moreover, we analysed how different censoring rates, with a focus on high censoring, affect the survival predictions of the model. We further explored the use of pseudo-observations as a way to handle censoring. Finally, we applied these methods to a breast cancer data set with large-scale gene expression data measured in tumours, in order to investigate the predictive ability of clinical and gene expression data for survival outcomes.

2   Overview of Existing Survival Methods

2.1   Notations for Survival Data

Let the random variable T denote the survival time. Let us assume that survival times and censoring times are independent and let us define the survival function as $S(t) = P(T > t)$. T_i is the realization of T for individual i, $1 \leq i \leq N$, with N the total number of individuals. Let D be the censoring index: if the realization of T for individual i is observed, $D_i = 1$ while if the corresponding data is right-censored, $D_i = 0$. We denote by $\tilde{T}$ the random variable such that $\tilde{T}_i = T_i$ if $D_i = 1$ and $\tilde{T}_i = \tau_i$, if $D_i = 0$, where τ_i is the censoring time for individual i. Thus we have: $\tilde{T}_i = \min(T_i, \tau_i)$. For every t, let us define $Z(t) = \mathbb{I}(T > t)$ and $Z_i(t)$ the realization of $Z(t)$ for individual i. Some of these realizations are observed, while some others are not, when $D_i = 0$ and $\tau_i \leq t$. Let $X = \{x_1, \ldots, x_p\}$ be a vector of covariates.

If all data were observed, an approximation of $S(t) = \mathbb{E}\left(Z(t)\right)$ would be given by the empirical mean $\frac{1}{N} \sum_{i=1}^{N} Z_i(t)$. With missing data, the Kaplan-Meier estimate computes the product of conditional probabilities (probability of surviving during a time interval provided you were alive at its beginning) and the information contained in the censored data is involved to compute the individuals at risk. We denote by $\hat{S}(t)$ the Kaplan-Meier estimate of $S(t)$.

2.2 CoxPH: the Cox Proportional Hazards Model

The reference model for survival analysis is the Cox proportional hazards model. It is set in continuous time. In this model, the relationship between the hazard function λ and the vector of covariates X takes the form:

$$\lambda(t|X_i) = \lambda_0(t)\exp(\phi(X_i)) \text{ with } \phi(X_i) = \beta^T X \tag{1}$$

The hazard function corresponds to the probability that an individual i will experience an event at time t, given that this individual has survived up to t. A key assumption of the model is that the hazard ratio for any two patients is constant over time.

2.3 Non-linear Survival Models Based on Neural Networks

Here, we explore different models that use multi-layer perceptron to perform survival analysis. These different models allow non-proportional effects. Various loss functions are used to fit each of them.

Kvamme et al. [10] introduce a method called Cox-CC that uses a special loss based on a case-control approximation. They propose to randomly sample a new set of controls at each iteration, instead of keeping control samples fixed.

$$\mathcal{L}_{\text{Cox-CC}} = \frac{1}{n} \sum_{i:D_i=1} \log\left(\sum_{j \in \tilde{R}^i} \exp[\phi(X_j) - \phi(X_i)]\right) \tag{2}$$

with $\tilde{R}^i$ a subset of the risk set R^i at time t including individual i. With their specific loss (Eq. (2)), the authors can fit a neural network using a mini-batch gradient descent algorithm.

They also present a second version of their model, CoxTime, that is not constrained by the proportionality assumption. To do so, they add the time variable as an additional input of the model. The loss function can then be rewritten as:

$$\mathcal{L}_{\text{Cox-Time}} = \frac{1}{n} \sum_{i:D_i=1} \log\left(\sum_{j \in \tilde{R}^i} \exp[\phi(t_i, X_j) - \phi(t_i, X_i)]\right) \tag{3}$$

Another proposed model, DeepHit [11], circumvents the proportional hazard assumption by using neural networks with a discrete time scale. A loss function is optimized that combines a log-likelihood with a ranking loss. We only study here the case of a single event. We define the time as: $0 = l_0 < \cdots < l_m$. $y(x) = [y_0(x), \ldots, y_m(x)]^T$ is the output of the neural network: given a patient i with covariates X_i, $y_s(X_i)$ is the probability that the patient will experience the event at time s. Thus we have:

$$\hat{S}(l_j|X_j) = 1 - \sum_{k=1}^{j} y_j(X) \tag{4}$$

The first term of the loss function is then written:

$$\mathcal{L}_{\text{DeepHit 1}} = -\frac{1}{n}\sum_{i=1}^{N} \underbrace{(D_i \log(y_{\kappa_i})}_{\text{Patients with event}} + \underbrace{(1 - D_i)\log(\hat{S}(\tilde{T}|X_i))}_{\text{Censored patients}} \tag{5}$$

κ_i is the index of the event time for individual i.

To the term in Eq. (5) is added another loss function which includes a constraint on the discrimination capacity of the model. It is an extension of the concordance index:

$$\mathcal{L}_{\text{DeepHit 2}} = \sum_{i,j} D_i \mathbb{I}\{\tilde{T}_i < \tilde{T}_j\}\exp(\frac{\hat{S}(\tilde{T}_i|X_i) - \hat{S}(\tilde{T}_i|X_j)}{\sigma}) \tag{6}$$

$\mathcal{L}_{\text{DeepHit 2}}$ (Eq. (6)) aims at improving the discrimination of the model by forcing the model to focus on times where there are a lot of events.

A third solution consists in using pseudo-observations. Andersen et al. [1] introduced pseudo-observations as an approach to perform regression based on the "jackknife" method. In survival analysis, the pseudo-observation for the individual i corresponds to the contribution of this individual to the Kaplan-Meier estimate $\hat{S}(t)$. Pseudo-observations are computed for all individuals at a given time-point, regardless of event time or censorship status.

It can be computed for different time points. For each individual i, we have k pseudo-observations. Let us define the pseudo-observation for individual i at time t:

$$\hat{Z}_i(t) = N\hat{S}(t) - (N-1)\hat{S}^{-i}(t) \tag{7}$$

Here, $\hat{S}^{-i}(t)$ denotes the leave-one-out Kaplan-Meier estimate $\hat{S}(t)$ obtained by discarding the data for individual i. We thus have an approximation of the empirical mean of the sample deprived from individual i

In their neural network model, DNNSurv, Zhao et al. [16] adapt the pseudo-observations approach in a discrete-time framework. Every survival time, censored or not, is replaced by the pseudo-observation computed as:

$$\hat{Z}_{ij}(t_{j+1}|R_j) = R_j\hat{S}(t_{j+1}|R_j) - (R_j - 1)\hat{S}^{-i}(t_{j+1}|R_j) \tag{8}$$

where $\hat{S}(t_{j+1}|R_j)$ is the Kaplan-Meier estimator constructed using the remaining survival times for all patients still at risk at time t_j, and $\hat{S}^{-i}(t_{j+1}|R_j)$ the Kaplan-Meier estimator for all patients at risk but the ith subject. The time variable is included in the input variables of the model. The loss function that is minimized boils down to the mean-squared-error.

3   Materials and Methods

3.1   Methods

In survival analysis, pseudo-observations provide a way to circumvent the complexity of censoring. We investigated here different ways of defining pseudo-observations.

In the work of Andersen et al. [1], pseudo-observations computation relies on the idea of leave-one-out estimation. Pseudo-observations approximate the missing values using the Kaplan-Meier estimate and can be computed as in Eq. (7). They are computed at a finite number of time points spread on the event time scale. The authors then use the pseudo-observations as the output of a generalized estimating equation model.

As in the DNNSurv model, we can also consider pseudo-observations in a discrete-time framework like we can see in Eq. (8). The time is divided in J intervals and pseudo conditional survival probabilities are computed for each interval. We implemented this method, called Pseudo-discrete. The discrete time variable is added as input of the neural network with the other covariates X. The predictions of the model correspond to the conditional survival probability in each interval. The marginal survival probability for the jth time is obtained by multiplying these conditional survival probabilities up to the jth interval.

An obvious problem with these methods is that it can take values strictly above 1 or below 0, which cannot be interpreted in a legitimate sense.

With the objective to interpret the pseudo-observations as the probability that $Z_i(t) = \mathbb{I}(T_i(t) > 0) = 1$, we enforced that it belongs to the $[0;1]$ interval, with the same idea used by classical pseudo-observations to take advantage of the information contained in the leave-one-out Kaplan-Meier estimators. Let us denote by $\bar{Z}(t) = (\bar{Z}_1(t), \cdots, \bar{Z}_N(t))$ the new vector of pseudo-observations. It is the solution of the constrained minimization problem:

$$\bar{Z}(t) = \arg \min_{u \in [0;1]^N} \|Au - b\|^2 \tag{9}$$

with A an $(N+1)$ by N matrix, and b an (N+1)-dimensional vectors given by:

$$A = \begin{pmatrix} \frac{1}{N} & \frac{1}{N} & \cdots & \frac{1}{N} \\ 0 & \frac{1}{N-1} & \cdots & \frac{1}{N-1} \\ \frac{1}{N-1} & 0 & \cdots & \frac{1}{N-1} \\ \vdots & \ddots & \ddots & \vdots \\ \frac{1}{N-1} & \cdots & \frac{1}{N-1} & 0 \end{pmatrix}, \quad b = \begin{pmatrix} \hat{S}(t) \\ \hat{S}^{-1}(t) \\ \vdots \\ \hat{S}^{-N}(t) \end{pmatrix} \tag{10}$$

Similarly as for the classical pseudo-observations, if there was no censoring, the solution of the problem would be $(Z_1(t), \cdots, Z_N(t))$. We call this new method Pseudo-optim. Again, we build a neural network including covariates and time variable as input. The output is the marginal survival probability for a given time.

Based on the idea that the pseudo-observation is an estimation of $Z_i(t)$, we can adopt a Bayesian point of view and take $\tilde{T}_i(t)$ as the conditional probability $\mathbb{P}(Z_i(t)|\mathcal{D})$, where $\mathcal{D}$ denotes the observed (uncomplete) data, $\mathcal{D} = (\tilde{T}_i, D_i)_{1 \leq i \leq N}$. If $D_i = 1$ or if $D_i = 0$ and $t < \tau_i$ (the censoring time for individual i), then $Z_i(t)$ is observed and we simply have: $\tilde{Z}_i(t) = Z_i(t)$. If $D_i = 0$ and $t \geq \tau_i$, then, since individual survivals are supposed independent and follow the same distribution law:

$$\mathbb{P}\left(Z_i(t) \mid \mathcal{D}\right) = \mathbb{P}\left(T_i > t \mid T_i > \tau_i, \mathcal{D}\right) = \mathbb{P}\left(T > t \mid T > \tau_i, \mathcal{D}\right) = \frac{\mathbb{P}\left(T > t \mid \mathcal{D}\right)}{\mathbb{P}\left(T > \tau_i, \mathcal{D}\right)}$$

(11)

Finally, $\mathbb{P}\left(T > t \mid \mathcal{D}\right)$ can be estimated by the Kaplan-Meier estimate, and thus, if $D_i = 0$ and $t \geq \tau_i$:

$$\tilde{Z}_i(t) = \frac{\hat{S}(t)}{\hat{S}(\tau_i)}$$

(12)

We call this the Pseudo-KM method.

3.2   Data

Simulation Study. In the simulation study, we further evaluated performance differences between the models according to the level of independent censoring in the data. We compared 2 censoring rates: 20% and 60%. For a given censoring rate, 100 data sets were generated. The data was simulated according to Friedman's random function generator [6,8]. It allowed us to generate random functions that have high-order interactions between the survival data and the explanatory variables and strong nonlinear effects. It is based on an Accelerated Failure Time (AFT) model, that assumes that the relationship between the logarithm of survival time T and the covariates is linear. The random function generator is defined as:

$$logT = m(X) + W \text{ with } W \sim \Gamma(2,1)$$

(13)

A vector of covariates $X = (X_1, ..., X_{20})$ is generated with $x \sim \mathcal{N}(0,1)$. $m(X)$ is defined by Friedman's random function generator:

$$m(X) = \sum_{l=1}^{10} a_l g_l(Z_l)$$

(14)

$\{a_l\}_1^{10}$ are randomly generated from a uniform distribution ($a_l \sim \mathcal{U}_{[-1,1]}$). R_l is a random subset of the input vector X of size n_l. The size of each subset, n_l, is itself random, with: $n_l = min(\lfloor 1.5 + r \rfloor, 10)$ and $r \sim \mathcal{E}(1/2)$. With Friedman's function generator, the input variables are associated with the survival time at different levels:

$$g_l(R_l) = \exp\{-\frac{1}{2}(R_l - \mu_l)^T V_l(R_l - \mu_l)\}$$

(15)

Each mean vector $\{\mu_l\}_1^{20}$ is randomly generated with $\mu_l \sim \mathcal{N}(0,1)$. The matrix of variance-covariance V_l is also randomly generated: $V_l = U_l D_l U_l^T$ with U_l an orthonormal random matrix, $D_l = diag\{d_{l,1}, ..., d_{l,n_l}\}$ and $\sqrt{d_{lk}} \sim \mathcal{U}(0.1, 2)$. We finally obtain the survival times by applying the exponential: $\exp(\log(T))$.

To simulate different levels of censoring in the data, we used an exponential law with parameter $\frac{1}{\gamma}\mathbf{E}(T)$. γ takes different values according to the censoring rate, with $\gamma = 4.5$ (resp. 0.9) for a censoring rate p of 20% (resp. 60%). The values of γ are obtained as a solution of $g(\gamma|p) = 0$ with $g(\gamma|p) = \mathbf{P}(D = 1|\gamma) - p$ and

$\mathbb{P}(D = 1|\gamma)$ the censoring probability in the population [15]. The minimum of the simulated survival time and the corresponding censoring time is taken.

METABRIC Data. The example data set we used is the METABRIC (Molecular Taxonomy of Breast Cancer International Consortium) cohort, which can be extracted from the MetaGxBreast package [7]. It consists of clinical features and large-scale gene expression data of breast cancer patients obtained at surgery. We used a nonspecific filter independent from outcome based on standard deviation to increase the statistical power [3], and retained the 863 genes with highest standard deviation. If multiple probe corresponded to the same gene, we selected the probe with the highest variance. Regarding the clinical covariates, we used the age at diagnosis, the grade, the tumor size, the number of invaded lymph nodes, hormonal therapy indicator and chemotherapy indicator. We removed individuals with missing values for the survival time. Since one or more covariate values were missing for 106 patients, we imputed the missing values using predictive mean matching for numerical features and a multinomial logit model for categorical variables [5]. Then, we standardized the numerical covariates using Z-score normalization and applied one-hot encoding to categorical variables. The final data set represents 1,960 patients and 869 features, with a median survival time of 88 months and a censoring rate of 54.6%.

3.3   Models Comparison

To determine the neural networks's hyperparameters, we compared different architectures. We tried out models with varying depths (number of hidden layers from 1 to 3) and neurons per layer ($\{4, 8, 32, 64, 128, 256, 514\}$). We considered 3 activation functions: Exponential Linear Unit, (ELU), Rectified Linear Units (ReLU) [4], and hyperbolic tangent (Tanh). Three optimization algorithms were tested: Adaptive Moment Estimation (Adam) [9], Root Mean Square Propagation (RMSProp) and Stochastic Gradient Descent with Warm Restarts (SGDWR) [12], with a learning rate varying from $1e-5$ to $1e-1$. Additionally, we applied regularization using a dropout layer at the end of each hidden layer of the network, with a rate from 0.1 to 0.5.

Then, we explored the hyperparameters search space by implementing a sequential model-based optimization using tree-structured Parzen estimators [2]. After building a surrogate model of the objective function, the algorithm allows us to select iteratively hyperparameters in an informed manner and update the surrogate model, using the Expected Improvement as criterion.

We evaluated the performances of the hyperparameters on the METABRIC data set using a double 5-fold cross validation [13] in order to mimick an external test set, as the data set is relatively small. First, the real data set was split into 5 folds: this is the outer loop. Then, we selected one of the 5 folds as test set and performed a 5-fold cross validation on the remaining data for each of the hyperparameters set: this is the inner loop. We chose the hyperparameters configuration with the minimum of the average validation loss obtained on the 5 folds of the inner-loop. Finally, we fitted the model with these optimal hyperparameters on the 4 folds of the outer loop and calculated predictions on the remaining test fold of the outer loop. This was repeated on all the folds of the outer loop.

For the simulation study, we generated a training set of size 2,000 and a test size of identical size. A 5-fold cross-validation was performed on the training set to select the set of parameters that minimize the average validation loss. Finally, we fitted the model with these optimal hyperparameters on the entire training set and calculated predictions on the test set.

3.4 Evaluation Criteria

The models were compared using 2 metrics: the concordance index for discrimination, and the Integrated Brier Score that measures both the discrimination and calibration of the model.

Concordance Index. The concordance index, or C-index, is a measure of the discrimination ability of a model, that is its ability to distinguish high-risk and low-risk patients. Specifically, it estimates the probability of agreement, i.e. the probability that two patients randomly selected are ordered in the same way between their survival prediction and their rank in the survival data. We used here a concordance measure proposed by Uno et al. [14] that accounts for the censored data using the inverse probability of censoring weighting. The concordance index for time t is then:

$$\hat{C}(t)_{IPCW} = \frac{\sum\limits_{i=1}^{N} \sum\limits_{j=1}^{N} D_i \hat{G}(\tilde{T}_i)^{-2} I\{\tilde{T}_i < \tilde{T}_j, \tilde{T}_i < t\} I\{\hat{S}(t|X_i) < \hat{S}(t|X_j)\}}{\sum\limits_{i=1}^{N} \sum\limits_{j=1}^{N} D_i \hat{G}(\tilde{T}_i)^{-2} I\{\tilde{T}_i < \tilde{T}_j, \tilde{T}_i < t\}} \quad (16)$$

where X is a $p \times 1$ covariate vector and $\hat{G}(.)$ is the Kaplan-Meier estimator for the censoring distribution $(G(t) = P(\tau > t))$ and I(.) is the indicator function. The value of the C-index lies between 0.5 and 1, with 0.5 representing a random prediction and 1 corresponding to a perfect ability to rank.

Brier Score. The Brier Score is used to evaluate the accuracy of the predicted survival function at a given time t. It is based on the root mean square error and focuses on the difference between the observed survival status and the predicted probability of survival. It lies between 0 (best possible value) and 1. The Brier Score for uncensored data is written as:

$$BS(t) = \frac{1}{N} \sum_{i=1}^{N} \left[\mathbb{I}\{T_i > t\} - \hat{S}(t|X_i) \right]^2$$
$$= \frac{1}{N} \sum_{i=1}^{N} \left[\hat{S}(t|X_i)^2 \mathbb{I}\{T_i \leq t\} + (1 - \hat{S}(t|X_i))^2 \mathbb{I}\{T_i > t\} \right] \quad (17)$$

With censored data, only a subset of the event times are observed. Graf et al. introduce a weighting of the Brier Score based on the inverse probability of censoring. It can be rewritten as:

$$\hat{BS}(t)_{IPCW} = \frac{1}{N} \sum_{i=1}^{N} \left[\hat{S}(t|X_i)^2 \frac{\mathbb{I}\{\tilde{T}_i \le t, D_i = 1\}}{\hat{G}(\tilde{T}_i)} \right.$$
$$\left. + (1 - \hat{S}(t|X_i))^2 \frac{\mathbb{I}\{\tilde{T}_i > t\}}{\hat{G}(\tilde{T}_i)} \right] \tag{18}$$

The Integrated Brier Score (IBS) is then written:

$$I\hat{B}S(t)_{IPCW} = \frac{1}{t_2 - t_1} \int_{t_1}^{t_2} BS(s)ds \tag{19}$$

4   Results and Discussion

4.1   Simulation Study

For each neural network method, we performed a 5-fold cross-validation on 100 different sets of parameters on the simulation data sets. Then we computed the C-index with a horizon fixed at the median survival time for the neural network trained on the finally chosen set of parameters. The Integrated Brier Score was computed from the 10th to the 90th percentile of the time distribution.

Table 1 represents the results for the median survival time for 20% and 60% censoring. All the models have better performances at 20% compared to 60% censoring. The best performing model in terms of numerically highest concordance indices at 20% censoring (resp. 60% censoring) is Cox-Time (resp. pseudo-discrete). Cox-CC performs better in terms of integrated Brier score with 60% censoring.

Table 1. Average C-index at median time and IBS for simulations with 20% and 60% censoring, and standard deviation across the 100 data sets. Bold values indicate highest values.

Censoring	C-index 20%	IBS 20%	C-index 60%	IBS 60%
CoxPH	0.734(±0.048)	0.147(±0.018)	0.533(±0.037)	0.185(±0.006)
Cox-CC	0.894(±0.027)	0.073(±0.013)	0.536(±0.036)	0.186(±0.007)
Cox-Time	**0.901(±0.02)**	**0.068(±0.01)**	0.536(±0.040)	**0.185(±0.008)**
DeepHit	0.845(±0.081)	0.214(±0.088)	0.530(±0.037)	0.207(±0.018)
Pseudo-disc	0.866(±0.054)	0.101(±0.035)	**0.540(±0.045)**	0.198(±0.064)
Pseudo-optim	0.891(±0.030)	0.085(±0.015)	0.519(±0.026)	0.224(±0.012)
Pseudo-KM	0.892(±0.028)	0.084(±0.016)	0.519(±0.027)	0.232(±0.012)

4.2   Example Data: METABRIC

For DeepHit, we defined the discrete time scale using 10 equally spaced time points. Furthermore, we computed pseudo-observations on the training set on equally distributed time points on the event time scale, up to the 80th percentile of survival distribution, that is at 22.01 years. We chose 80th percentile as upper bound to ensure enough subjects remain at risk in the study. We only computed the values until an event or censoring occurs for discrete pseudo-observations (Pseudo-disc), as is done in the original paper of DNNSurv.

All comparison measures were computed on the test set, at 5 and 10 years. These times are chosen as these are usual times of interest for the clinical investigators for early breast. In terms of C-indices, we see on Table 2 that the best performing model is the Pseudo-optim model. The numerically highest value for the integrated Brier score is obtained by DeepHit. All the models obtain better discrimination at 5 years, compared to 10 years.

Table 2. Average C-index and Integrated Brier Score with standard deviation across the 5 folds for METABRIC data. Highest values are in bold.

	C-index 5 years	C-index 10 years	IBS
CoxPH	0.568(±0.034)	0.549(±0.032)	0.363(±0.054)
Cox-CC	0.667(±0.032)	0.640(±0.024)	0.202(±0.002)
Cox-Time	0.670(±0.017)	0.638(±0.024)	0.205(±0.022)
DeepHit	0.673(±0.031)	0.619(±0.015)	**0.195(±0.022)**
Pseudo-disc	0.674(±0.016)	0.637(±0.022)	0.204(±0.022)
Pseudo-optim	**0.683(±0.029)**	**0.643(±0.035)**	0.206(±0.022)
Pseudo-KM	0.665(±0.025)	0.623(±0.022)	0.214(±0.019)

5   Discussion and Conclusion

We presented here a comparison of different models based on neural networks to perform survival analysis. These models are not constrained by the proportionality assumption of the CoxPH model and are based on specific loss functions. In addition, we focused on the use of pseudo-observations to deal with censored observations and explored different ways of computing them, either in a discrete or continuous time, with a $[0, 1]$ constraint and in a Bayesian view. The constraint has the advantage to see the pseudo-observation as an individual's survival probability at a given time point.

As artificial neural networks can model complex relationships between the covariates and event times, we simulated data with interactions and nonlinearity using a random function generator based on an AFT model, and with different levels of censoring. We observed that many of the proposed neural network models had overall rather similar performances. We demonstrated numerically superior results of CoxTime at lower censorship and Pseudo-disc at higher censorship in terms of concordance index. However, the standard deviation is quite wide and cover different methods. At 20% censorship, all the neural networks performed better than the linear model and fitted to the complex interactions existing in the simulated data. There was a drop in performance for all the models at higher censorship.

In the example data of early breast cancer, we obtained overall good performances, especially at 5 years. Furthermore, Pseudo-optim had slightly better discrimination performances than the other models. The Cox model with linear effects underperformed compared to artificial neural networks.

In future research, we could explore the value of these neural network methods in the context of higher-dimensional data and thus further raise the number of covariates, but this may need larger sample sizes. We could also challenge our results by simulating data differently and use other methods as a benchmark, like random survival forest models.

References

1. Andersen, P.K., Pohar Perme, M.: Pseudo-observations in survival analysis, February 2010. https://doi.org/10.1177/0962280209105020
2. Bergstra, J., Bardenet, R., Bengio, Y., Kégl, B.: Algorithms for hyper-parameter optimization. In: Advances in Neural Information Processing Systems 24: 25th Annual Conference on Neural Information Processing Systems 2011, NIPS 2011, pp. 1–9 (2011)
3. Bourgon, R., Gentleman, R., Huber, W.: Independent filtering increases detection power for high-throughput experiments. Proc. Natl. Acad. Sci. U.S.A. **107**(21), 9546–9551 (2010). https://doi.org/10.1073/pnas.0914005107
4. Brown, M.J., Hutchinson, L.A., Rainbow, M.J., Deluzio, K.J., De Asha, A.R.: A comparison of self-selected walking speeds and walking speed variability when data are collected during repeated discrete trials and during continuous walking. J. Appl. Biomech. **33**(5), 384–387 (2017). https://doi.org/10.1123/jab.2016-0355
5. van Buuren, S., Groothuis-Oudshoorn, K.: mice: Multivariate imputation by chained equations in R. J. Stat. Softw. **45**(3), 1–67 (2011). https://doi.org/10.18637/jss.v045.i03
6. Friedman, J.H.: Greedy function approximation: a gradient boosting machine. Ann. Stat. **29**(5), 1189–1232 (2001). https://doi.org/10.2307/2699986
7. Gendoo, D.M., et al.: MetaGxData: clinically annotated breast, ovarian and pancreatic cancer datasets and their use in generating a multi-cancer gene signature. Sci. Rep. **9**(1), 1–14 (2019). https://doi.org/10.1038/s41598-019-45165-4
8. Henderson, N.C., Louis, T.A., Rosner, G.L., Varadhan, R.: Individualized treatment effects with censored data via fully nonparametric Bayesian accelerated failure time models. Biostatistics **21**(1), 50–68 (2020). https://doi.org/10.1093/biostatistics/kxy028
9. Kingma, D.P., Ba, J.L.: Adam: a method for stochastic optimization. In: 3rd International Conference on Learning Representations, ICLR 2015 - Conference Track Proceedings, pp. 1–15 (2015)
10. Kvamme, H., Borgan, Ø., Scheel, I.: Time-to-event prediction with neural networks and cox regression. J. Mach. Learn. Res. **20** (2019). http://arxiv.org/abs/1907.00825

11. Lee, C., Zame, W.R., Yoon, J., Van Der Schaar, M.: DeepHit: a deep learning approach to survival analysis with competing risks. In: 32nd AAAI Conference on Artificial Intelligence, AAAI 2018, pp. 2314–2321 (2018)
12. Loshchilov, I., Hutter, F.: SGDR: stochastic gradient descent with warm restarts. In: 5th International Conference on Learning Representations, ICLR 2017 - Conference Track Proceedings, pp. 1–16 (2017)
13. Simon, R., Subramanian, J., Li, M.C., Menezes, S.: Using cross-validation to evaluate predictive accuracy of survival risk classifiers based on high-dimensional data. Brief. Bioinform. **12**(3), 203–14 (2011)
14. Uno, H., Cai, T., Pencina, M.J., D'Agostino, R.B., Wei, L.J.: On the C-statistics for evaluating overall adequacy of risk prediction procedures with censored survival data. Stat. Med. **30**(10), 1105–1117 (2011). https://doi.org/10.1002/sim.4154. https://onlinelibrary.wiley.com/doi/abs/10.1002/sim.4154
15. Wan, F.: Simulating survival data with predefined censoring rates for proportional hazards models. Stat. Med. **36**(5), 838–854 (2017). https://doi.org/10.1002/sim.7178
16. Zhao, L., Feng, D.: DNNSurv: Deep Neural Networks for Survival Analysis Using Pseudo Values, pp. 1–15 (2019). http://arxiv.org/abs/1908.02337

Mathematical Modeling for Cancer Research

tughall: A Tool to Reproduce Darwinian Evolution of Cancer Cells for Simulation-Based Personalized Medicine

Iurii Nagornov$^{(\boxtimes)}$, Jo Nishino, and Mamoru Kato

Division of Bioinformatics, National Cancer Center Japan, 5-1-1 Tsukiji, Chuo-ku, Tokyo 104-0045, Japan
inagonov@ncc.go.jp

Abstract. Here, we present new version 2.1 of **tugHall** *(tumor gene-Hallmark)* cancer-cell evolution simulator which is accelerated by clone-based approach. The tool is based on the model connected to the well-known cancer hallmarks with the specific mutational states of tumor-related genes. The hallmark variables depend linearly on the mutational states of tumor-related genes with specific weights. The cell behavior phenotypes are stochastically determined and the phenotypic probabilities are probabilistically interfered by the hallmarks. Approximate Bayesian Computation is applied to find the personalized specific parameters of the model. The variant allele frequencies are used as target data for the analysis. In tugHall 2.1, the Darwinian evolutionary competition amongst different clones is computed due to clone's death/birth processes. The open-source code is available in the repository www.github.com/tugHall.

Keywords: Cancer cell evolution · Cancer hallmarks · Approximate Bayesian Computation

1 Model

We recently released version **tugHall 2.1**. Briefly (check [1] for more details), in **tugHall**, cells at an initial timepoint are put on trials, where the next phenotypic state of each cell is probabilistically determined. Cancer hallmarks are introduced as interfering factors during the probabilistic determination of the phenotypic cascade. For example, during the "apoptosis" trial wherein the cell is destined to undergo "apoptosis" with a probability variable, a, the "evading apoptosis" or shortly "apoptosis" hallmark (H_a) [2], is introduced. This decreases the apoptosis probability by $a - H_a$. The cancer hallmark (H) variables depend linearly on the combined variables of mutational states and their constant weights [1]. tugHall is thus an agent-based model of branching cascades to stochastically determine destined cellular phenotype in the presence of weighted H as interfering factors. Unlike versions 1.0 and 1.1 which are cell-based, **tugHall 2.1** is a clone-based code. In this program a clone is defined as a set of cells with the exact same genomic (mutational) states. Thus, grouping genetically the same cells together brings down the number of cells whose phenotype is to be determined. This accelerates

© Springer Nature Switzerland AG 2020
G. Bebis et al. (Eds.): ISMCO 2020, LNBI 12508, pp. 71–76, 2020.
https://doi.org/10.1007/978-3-030-64511-3_7

the computation speed of the algorithm. Moreover, **version 2.1** has additional changes in the parts concerning binominal distribution which allows calculation of binominal distribution for numbers $n \geq 10^{14}$. For such large numbers we used approximation with normal distribution, because function *rbinom*() has a limitation for numbers. With this modification, the computation acceleration rate observed is about 10^3 to 10^4 for a single clone or several clones with 1000 cells.

The dependencies in the algorithm are represented by several levels:

- First level is represented by mutation rate for each gene m', which equals the product of constant m_0 and the length of CDS. A mutation generated under the mutation rate can be a driver mutation with probability u' (u_o for oncogenes and u_s for suppressors) to mutate a tumor-related gene; otherwise, a passenger mutation, which has no effects on a gene and just increases the number of mutations.
- Second level relates to the linear dependences of hallmarks (H_a, H_i, H_{im}, H_d, H_b – apoptosis, immortality, invasion/metastasis, oncogene/suppressor and angiogenesis hallmarks, respectively) on mutated genes with related weights w_a, w_i, w_{im}, w_d, w_b. For example:

$$H_a = \sum_{k=1}^{N_g} w_a^k G_k, \ w_a = w_a^k,$$

where N_g is a number of genes, G_k is a binary indicator of mutated gene k.
- Third level includes the probabilities for all cell-behavior processes, which are functions of correspondent hallmarks [1].

2 Software

The open-source code of tugHall is available in two versions, first is cell-based code and second is clone-based code. The repository is deposited at https://github.com/tugHall. It includes documentation, tests and results of the testing. The analysis script allows to calculate the following information:

1) The temporal evolution of cell number for primary tumor and metastatic cells
2) The temporal evolution of the hallmarks and their probabilities
3) The number of cells vs clones defined by identical driver and all (driver plus passenger) mutated genes.
4) A histogram for inequality (Gini) coefficients based on the number of cells in clones.
5) The distribution of impaired genes and the order of the gene's dysfunction.

3 Simulations of Darwinian Evolution Based on Personalized Weighting of Hallmarks for Two Cancer Patients

Data deposited in global resources such as the International Cancer Genome Consortium (ICGC) [3] and The Cancer Genome Atlas (TCGA) provide variant allele frequencies (VAF) data on tumor-related genes. We extracted VAFs from TCGA data base (Table 1) for two colorectal cancer patients with ID TCGA-AF-5654-01A-01D-1657-10 and TCGA-D5-5540-01A-01D-1650-10 (TCGA-AF-* and TCGA-D5-* in short, respectively). Simulation parameters were estimated using the approximate Bayesian computation (ABC) framework. Previous studies have successfully used an ABC method in a branching process model, making parameter estimates using summary statistics such as VAFs in bulk-cell signaling data [1, 4]. We used COSMIC [5] to limit the range of the values of weights for hallmarks.

In order to estimate weights $(w_x)_k$ for these patients, we used 10000 parallel independent calculations. The calculations were conducted by assigning random values taken

Table 1. VAF for two patients from TCGA data base

ID in TCGA data base	APC	KRAS	TP53	PIK3CA
TCGA-AF-5654-01A-01D-1657-10	0.4471545	0.5988372	0.0	0.0
TCGA-D5-5540-01A-01D-1650-10	0.4713896	0.4910394	0.7826087	0.8588235

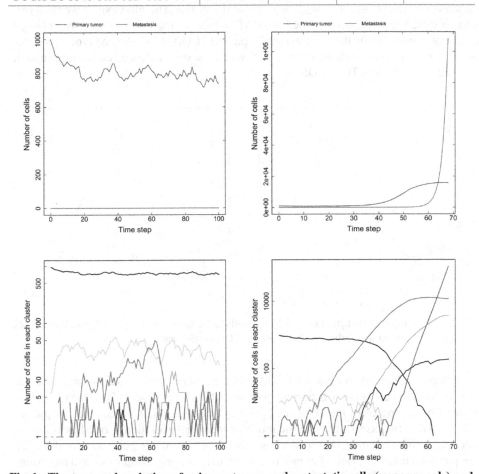

Fig. 1. **The temporal evolution of primary tumor and metastatic cells (upper panels) and the number of cells in the clusters (lower panels).** Left panels represent **TCGA-AF-*** patient and rights panel represent **TCGA-D5-*** patient.

from uniform distribution priors to weights. VAF data from Table 1 are used as summary statistics for ABC to estimate the posterior distributions of the weight parameters. Subsequently, MAP (maximum a posteriori) estimators were calculated for each patient. These personalized weights $(w_x)_k$ of the model were used for further iterations (600 trials for each patient with fixed weights with the MAP values).

The computation results for the two patients are shown in Figs. 1, 2 and 3. Figure 1 shows typical evolution of primary tumor and metastatic cells. During the progress of a trial, clusters of clone cells with the same set of **driver** mutations evolve. Further, Fig. 1 shows Darwinian evolution (competition between clusters); during evolution many clusters died out while few escaped extinctions.

Since the number of clones is much larger than the number of clusters, we focused on the first 100 clones. Figure 2 shows that most of clusters died within the first two steps of their genesis. There were a few clones that survived during Darwinian evolution. The most interesting feature of tugHall is that the survival evolution crucially depends on weights w_a, w_i, w_{im}, w_d, w_b in the model. Therefore, since we used two sets of different weights calculated as MAP values for the two patients, we obtained different results. While the simulations for patient TCGA-AF-* showed typical competition between clones without any resultant winner, several extended clones appeared in simulation for patient TCGA-D5-*.

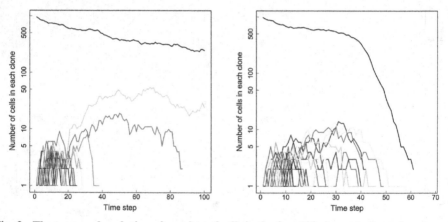

Fig. 2. The temporal evolution of number of cells in the first 100 clones. Left panel represents **TCGA-AF-*** patient and rights panel represents **TCGA-D5-*** patient.

Interestingly, while for patient TCGA-AF-* the metastatic cells did not appear in all simulations, the converse was true for patient TCGA-D5-* (Fig. 3) [1]. Due to appearance of metastatic cells, the number of primary tumor cells is less than that for patient TCGA-AF-*. In accordance with the model, cancer cell evolution is ruled by relations between several factors. Death of clones is due to apoptosis but neutralized by the apoptosis hallmark; growth of clones is through the oncogene/suppressor and angiogenesis hallmarks while death occurs due to the environmental death represented by the parameter k [1]. Each factor represents the sum of the influences of genes with weights

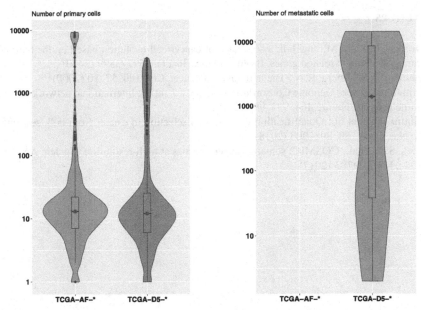

Fig. 3. **The number of primary tumor and metastatic cells** for repeated simulations for patients **TCGA-D5-*** and **TCGA-AF-***.

w_a, w_i, w_{im}, w_d, w_b. Moreover, the weights define the behaviors of clones that contain the driver gene mutations acquired during the progression of trial simulation. Thus, complexity of simulations lies not only in the multi-dimensional space of parameters, but also in the stochastic nature of Markov chain model.

4   Conclusion

tugHall is based on a stochastic model of cancer-cell evolution and includes the hallmarks of cancer. The effects of hallmarks on cell behaviors during the course of the simulation change due to mutations of genes. These influences of mutated genes on hallmarks are defined as linear functions of weights (w_a, w_i, w_{im}, w_d, w_b) and the mutational states of tumor-related genes. Evolution is ruled by relations between factors governing death and genesis. While cellular death is brought about by the apoptosis and environmental factors, the growth of clone is governed by the oncogene/suppressor and angiogenesis hallmarks. In principle it is possible to find the personalized weights, which result in different evolutionary scenarios, using ABC method as demonstrated in the two representative cases of cancer patients.

Acknowledgements. We are grateful to Momoko Nagai and Atsushi Niida for fruitful discussion and useful suggestions.

Funding. This work was supported by JST CREST [14531766]; MEXT [20K12071 and 19H03440].

References

1. Nagornov, I., Kato, M.: tugHall: a simulator of cancer-cell evolution based on the hallmarks of cancer and tumor-related genes. Bioinformatics **36**(11), 3597–3599 (2020)
2. Hanahan, D., Weinberg, R.A.: The hallmarks of cancer. Cell **100**, 57–70 (2000)
3. International Cancer Genome Consortium, Hudson, T.J., et al.: International network of cancer genome projects. Nature **464**, 993–998 (2010)
4. Williams, M.J., et al.: Quantification of subclonal selection in cancer from bulk sequencing data. Nat. Genet. **50**, 895–903 (2018)
5. Forbes, S.A., et al.: COSMIC: somatic cancer genetics at high-resolution. Nucleic Acids Res. **45**(D1), D777–D783 (2017)

General Cancer Computational Biology

The Potential of Single Cell RNA-Sequencing Data for the Prediction of Gastric Cancer Serum Biomarkers

Kirill E. Medvedev[1](✉), Anna V. Savelyeva[2], Aditya Bagrodia[2], and Nick V. Grishin[1,3,4]

[1] Department of Biophysics, University of Texas Southwestern Medical Center, Dallas, TX, USA
`Kirill.Medvedev@UTSouthwestern.edu`
[2] Department of Urology, University of Texas Southwestern Medical Center, Dallas, TX, USA
[3] Department of Biochemistry, University of Texas Southwestern Medical Center, Dallas, TX, USA
[4] Howard Hughes Medical Institute, University of Texas Southwestern Medical Center, Dallas, TX, USA

Abstract. Gastric cancer (GC) is the sixth most common worldwide malignancy and the third leading cancer cause of death. Early diagnosis and effective after-surgical monitoring can significantly improve survival rates. Previous studies have revealed several serum biomarkers that are elevated in GC patients, including CEA, CA19-9, and CA72-4. However, sensitivity of these biomarkers is below 30%. Identification of more sensitive and specific to GC markers is critical for individualized therapy of this disease. Here we developed an approach for single-cell transcriptomic data analysis that identifies secretory proteins that are abundantly expressed in GC cells and that could be measurable in the blood. Using early GC scRNA-seq data, we identified 19 secretory proteins significantly overexpressed in GC cells. Notably, 4 proteins (IL32, KLK10, KLK7, OLFM4) have demonstrated more superior sensitivity in comparison to conventional serum markers in previous studies. Moreover, 2 proteins, F12 and CFD, were not previously associated with GC and were not utilized for serum-based testing of other malignancies. Proposed approach has a high potential to be used for serum marker identification in other types of cancers and presented here data could be a source for the development of more sensitive and specific diagnostic panel for early gastric cancer detection and patient post-treatment monitoring.

Keywords: Gastric cancer · Gastric cancer secretome · Biomarkers of gastric cancer

1 Introduction

Gastric (stomach) cancer (GC) remains one of the most common and most lethal cancers worldwide [1]. Despite recent successes in the treatment and improved survival

---

K. E. Medvedev and A. V. Savelyeva—These authors contributed equally to this work.

© Springer Nature Switzerland AG 2020
G. Bebis et al. (Eds.): ISMCO 2020, LNBI 12508, pp. 79–84, 2020.
https://doi.org/10.1007/978-3-030-64511-3_8

of patients, GC is often diagnosed at later stages with associated poor prognosis [2]. Because the progression of the GC is mostly asymptomatic until the latest stages, methods for early detection of this disease using less-invasive screening approaches are in great demand. Therefore, there is critical unmet need for new GC serum biomarkers that can be used for patient screening and monitoring. Sensitivity of currently used in clinical labs biomarkers for GC diagnostic (CEA, CA19-9, and CA72-4) is below 30% [3]. Here we use single cell RNA-sequencing data to identify new potentially secreted biomarkers of GC cells, which may be used for the development of more specific and sensitive screening approach.

2 Materials and Methods

To characterize the GC secretome, we used publicly available single cell RNA-seq data of an early intestinal gastric cancer sample (EGC) and three samples of non-atrophic gastritis (NAG) collected from three different patients. NAG is the initial premalignant stage of gastric cancer that was used as a sample of comparison to identify secretory proteins abundant in oncotransformed GC cells at the early stages but not characteristic for premalignant initial stages of this disease [4]. Totally 10,830 cells were included in the pipeline. We processed scRNA-seq raw data by the Cell Ranger Single-Cell Software Suite (release 2.0), including using cellranger count to perform alignment, filtering, barcode counting, and UMI counting. The reads were aligned to the hg19 reference genome using a 10X Genomics pre-built annotation package. The output from different samples was aggregated using cellranger aggr with default parameter setting. To find specific biomarkers secreted by GC cells, we first identified cancer cells cluster (CCC) in EGC sample using well-known gastric cancer markers CEACAM5 (CEA) and CEACAM6, as also suggested by Zhang et al. [4]. To define distinctive features of cancer cells, we compared genes level of expression between CCC and pit mucous cells cluster (PMC), defined by markers MUC5AC and TFF1. Pit mucous cells is the most common cell type of gastric mucosa that is known to be an initial site of precancerous lesion formation. From the top 200 significantly upregulated genes in CCC we filtered 19 secretory proteins using UniProt Knowledge Base [5]. Although the level of gene expression may not correlate with protein secretion level, this analysis allows to identify the most promising candidates for the experimental validation. To estimate and compare the predictive power for 19 secretory proteins elevated in GC cells, we evaluated ROC curves of their gene expression levels. We've also used expression level of well-known gastric cancer biomarkers (CEA, CA19-9, and CA72-4) as a positive reference control. False Positive Rate (FPR) and True Positive Rate (TPR) for ROC curves were calculated for the EGC sample versus all NAG samples. True Positives (TP) for particular gene were calculated as a number of cells in cancer sample with Log2 of expression level at a particular cutoff. We used Log2(gene expression level) > 0 as the lowest cutoff and (Log2(gene expression level) > maximum of expression level for particular gene) as the highest cutoff. We calculated False Positives (FP), True Negatives (TN) and False Negatives (FN) similarly. We calculated the area under the ROC curve (AUC) for each of 19 genes.

3 Results

Analysis of gene expression levels and cells clustering using tSNE algorithm revealed significant differences between EGC and NAG samples (Fig. 1A). Identified CCC and PMC clusters (Fig. 1B) were compared and, from the top 200 significantly upregulated genes in CCC, 19 secretory proteins that could serve as serum biomarkers for patients with early intestinal GC were selected. Importantly, 17 of the proteins (PRAP1 [6], OLFM4 [7], KLK10, KLK7 [4], SERPINA1 [8], REG1A [9], REG4 [10], KLK6 [11], CCL25 [12], MUC17 [13], CEMIP [14], PLA2G2A [15], IL32 [16], CCL15 [17], ODAM [18], PRSS3 [19], MSLN [20]) were previously described as prospective markers for

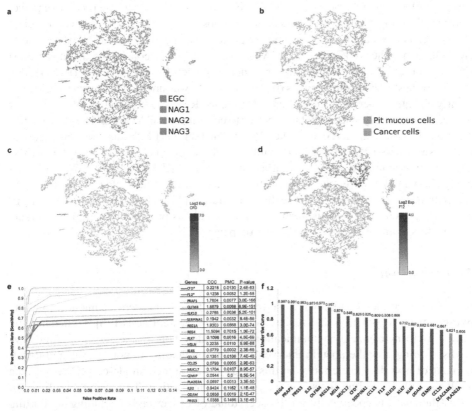

Fig. 1. The tSNE plot of clustered cells from four samples, where (A) EGC stands for early gastric cancer, NAG - non-atrophic gastritis; (B) pit mucous cells were defined using MUC5AC and TFF1 markers, cancers cells using CEACAM5 and CEACAM6; (C) color shows expression of CFD; (D) color shows expression of F12; (E) ROC curves plot of top upregulated secretory proteins in gastric cancer cluster. Average level of expression for selected genes in CCC and PMC clusters is shown in the legend. Genes, which we suggested as potential biomarkers denoted by asterisk and thick lines. (F) The Area Under de Curve (AUC) for the Receiver Operating Characteristic (ROC) curves for top 19 upregulated genes in cancer cells cluster. Genes, which we suggested as potential biomarkers denoted by asterisk and colored in red. CEACAM5, a well-known GC biomarker, used as a positive reference control (colored in green). (Color figure online)

diagnostic histopathology of GC. Diagnostic potential of two remaining proteins (CFD and F12) was not previously explored for this disease. Notably, CFD and F12 are known to be involved in tumor progression and microenvironment formation of other cancer types. For example, Complement factor D (CFD), or adipsin, was shown to enhance breast cancer cells growth via cancer stem cells niche formation [21]. Coagulation factor XII (F12) is involved in tumor vascular permeability [22]. Significant overexpression of CFD and F12 in CCC (Fig. 1C and 1D respectively), suggest these proteins as new promising biomarkers for gastric cancer diagnostic.

We used well-known gastric cancer biomarkers (CEA, CA19-9, and CA72-4) as a positive reference control; however, only CEA was detected in the majority of cancer cells and could be utilized for this analysis. The area under curve (AUC) for the Receiver Operating Characteristic (ROC) (Fig. 1E and 1F) reflects the predictive power of biomarkers identified in analyzed data. The closer the AUC is to 1, the more efficient biomarker is expected to be. 18 secretory proteins with elevated level of expression in GC cells had higher AUC than CEACAM5 (CEA), which is considered the most robust GC biomarker [23, 24]. Importantly, regenerating islet-derived protein 4 (REG4), that has the highest AUC in our analysis (0.997), (Fig. 1F) was previously shown to be significantly elevated in serum of GC patients. Moreover, REG4 demonstrated 73% sensitivity for GC (even at early stages) that is superior to conventional CA19-9 and CEA tests [10]. The fourth (based on our AUC-ranked list) biomarker, OLFM4, has also demonstrated higher sensitivity (25%) in patients with stage I GC than CA19-9 (5%) or CEA (3%) [25]. Another potential secretory marker from our list, KLK6 (AUC 0.697) was previously shown to be up-regulated in GC tissues and has 1.7-fold higher concentration in GC patient sera in comparison to healthy individuals [11]. Therefore, we suggest all 19 secretory proteins of GC cells identified in this study as prospective serum biomarkers for the development of early GC diagnostic panel.

4 Discussion

Cancer cell secretome is known to be involved in tumor microenvironment formation, malignant transformation and maintenance of tumor stem cells [26]. Understanding tumor microenvironment communications is considered one of the most important keys to discovering new therapeutic interventions, understanding tumorigenesis and development of new diagnostic tools. In this study we combined single-cell RNA-seq data analysis with a pipeline for secretory protein identification. This approach allows to focus the analysis specifically on the secretome of cancer cells excluding supporting tissues from the investigation. Gastric cancer is one of the worldwide most common malignancies with the high rate of death due to the belated diagnosis. Widely acknowledged GC biomarkers including CEA, CA19-9, and CA72-4 have a very low sensitivity level (30%) [3] that could be improved by either expansion of the biomarker panel or substitution of these markers with more reliable ones. We showed that gastric cancer cells in comparison to normal pit mucous cells significantly overexpress 19 secretory proteins. Notably, 2 proteins (REG4 and OLFM4) have demonstrated more superior sensitivity in comparison to conventional serum markers (CEA and CA19-9) in previous studies [10, 25]. Moreover, 17 other markers were not previously verified for GC serum diagnostics and 8 of them (CCL14, F12, CFD, PRAP1, ODAM, PLA2G2A, PRSS3, MUC17) were not proposed as serum biomarkers for any other cancer type and could potentially represent GC-specific pool of diagnostic markers.

Identification of serum biomarkers for early GC detection will have a tremendous impact on GC diagnosis, treatment, and monitoring. Our data encompass a great source for the development of more sensitive and specific diagnostic panel for stage I-II gastric cancer detection and post-treatment monitoring of GC patients.

References

1. Jemal, A., Bray, F., Center, M.M., Ferlay, J., Ward, E., Forman, D.: Global cancer statistics. CA Cancer J. Clin. **61**(2), 69–90 (2011)
2. Matsuoka, T., Yashiro, M.: Biomarkers of gastric cancer: current topics and future perspective. World J. Gastroenterol. **24**(26), 2818 (2018)
3. Shimada, H., Noie, T., Ohashi, M., Oba, K., Takahashi, Y.: Clinical significance of serum tumor markers for gastric cancer: a systematic review of literature by the Task Force of the Japanese Gastric Cancer Association. Gastric Cancer **17**(1), 26–33 (2013). https://doi.org/10.1007/s10120-013-0259-5
4. Zhang, P., et al.: Dissecting the single-cell transcriptome network underlying gastric premalignant lesions and early gastric cancer. Cell Rep. **27**(6), 1934–1947 (2019)
5. UniProt Consortium: UniProt: a worldwide hub of protein knowledge. Nucleic Acids Res. **47**(D1), D506–D515 (2019)
6. Gu, W., et al.: LncRNA expression profile reveals the potential role of lncRNAs in gastric carcinogenesis. Cancer Biomark. **15**(3), 249–258 (2015)
7. Luo, Z., Zhang, Q., Zhao, Z., Li, B., Chen, J., Wang, Y.: OLFM4 is associated with lymph node metastasis and poor prognosis in patients with gastric cancer. J. Cancer Res. Clin. Oncol. **137**(11), 1713 (2011). https://doi.org/10.1007/s00432-011-1042-9
8. Kwon, C.H., et al.: Serpin peptidase inhibitor clade A member 1 is a biomarker of poor prognosis in gastric cancer. Br. J. Cancer **111**(10), 1993–2002 (2014)
9. Qiu, Y.S., Liao, G.J., Jiang, N.N.: DNA methylation-mediated silencing of regenerating protein 1 Alpha (REG1A) Affects Gastric Cancer Prognosis. Med. Sci. Monit. **23**, 5834 (2017)
10. Kobayashi, Y., et al.: Serum tumor antigen REG4 as a useful diagnostic biomarker in gastric cancer. Hepatogastroenterology **57**(104), 1631–1634 (2010)
11. Kim, J.J., et al.: Upregulation and secretion of kallikrein-related peptidase 6 (KLK6) in gastric cancer. Tumor Biol. **33**(3), 731–738 (2012). https://doi.org/10.1007/s13277-011-0267-1
12. Lillard, J.W., Singh, R., Singh, S.: Detecting cancer with anti-CCL25 and anti-CCR9 antibodies. Morehouse School of Medicine Inc, assignee. United States patent US 8,658,377 (2014)
13. Lin, S., et al.: Epigenetic downregulation of MUC17 by H. pylori infection facilitates NF-κB-mediated expression of CEACAM1-3S in human gastric cancer. Gastric Cancer **22**(5), 941–954 (2019)
14. Jia, S., et al.: KIAA1199 promotes migration and invasion by Wnt/β-catenin pathway and MMPs mediated EMT progression and serves as a poor prognosis marker in gastric cancer. PLoS One **12**(4), e0175058 (2017)
15. Ganesan, K., et al.: Inhibition of gastric cancer invasion and metastasis by PLA2G2A, a novel β-catenin/TCF target gene. Cancer Res. **68**(11), 4277–4286 (2008)
16. Ishigami, S., et al.: IL-32 expression is an independent prognostic marker for gastric cancer. Med. Oncol. **30**(2), 472 (2013). https://doi.org/10.1007/s12032-013-0472-4

17. Raja, U.M., Gopal, G., Shirley, S., Ramakrishnan, A.S., Rajkumar, T.: Immunohistochemical expression and localization of cytokines/chemokines/growth factors in gastric cancer. Cytokine **89**, 82–90 (2017)
18. Aung, P.P., et al.: Systematic search for gastric cancer-specific genes based on SAGE data: melanoma inhibitory activity and matrix metalloproteinase-10 are novel prognostic factors in patients with gastric cancer. Oncogene **25**(17), 2546–2557 (2006)
19. Wang, F., et al.: High-level expression of PRSS3 correlates with metastasis and poor prognosis in patients with gastric cancer. J. Surg. Oncol. **119**(8), 1108–1121 (2019)
20. Einama, T., et al.: Luminal membrane expression of mesothelin is a prominent poor prognostic factor for gastric cancer. Br. J. Cancer **107**(1), 137–142 (2012)
21. Goto, H., et al.: Adipose-derived stem cells enhance human breast cancer growth and cancer stem cell-like properties through adipsin. Oncogene **38**(6), 767–779 (2019)
22. Matsumura, Y.: Cancer stromal targeting (CAST) therapy. Adv. Drug Deliv. Rev. **64**(8), 710–719 (2012)
23. He, C.Z., Zhang, K.H., Li, Q., Liu, X.H., Hong, Y., Lv, N.H.: Combined use of AFP, CEA, CA125 and CA19-9 improves the sensitivity for the diagnosis of gastric cancer. BMC Gastroenterol. **13**(1), 1–5 (2013)
24. Marrelli, D., et al.: Prognostic significance of CEA, CA 19-9 and CA 72-4 preoperative serum levels in gastric carcinoma. Oncology. **57**(1), 55–62 (1999)
25. Oue, N., et al.: Serum olfactomedin 4 (GW112, hGC-1) in combination with Reg IV is a highly sensitive biomarker for gastric cancer patients. Int. J. Cancer **125**(10), 2383–2392 (2009)
26. Paltridge, J.L., Belle, L., Khew-Goodall, Y.: The secretome in cancer progression. Biochim. Biophys. Acta **1834**(11), 2233–2241 (2013)

Posters

Theoretical Foundation of the Performance of Phylogeny-Based Somatic Variant Detection

Takuya Moriyama[1], Seiya Imoto[1], Satoru Miyano[2], and Rui Yamaguchi[1,3,4(✉)]

[1] Human Genome Center, The Institute of Medical Science, The University of Tokyo, Tokyo, Japan
moriyama@hgc.jp, imoto@ims.u-tokyo.ac.jp
[2] M&D Data Science Center, Tokyo Medical and Dental University, Tokyo, Japan
miyano@hgc.jp
[3] Division of Cancer Systems Biology, Aichi Cancer Center Research Institute, Nagoya, Japan
r.yamaguchi@aichi-cc.jp
[4] Department of Cancer Informatics, Nagoya University Graduate School of Medicine, Nagoya, Japan

Abstract. We study the performance of a variant detection method that is based on a property of tumor phylogenetic tree. Our major contributions are two folds. First, we show the property of tumor phylogenetic tree: the total patterns of mutations are restricted if a multi-regional mutation profile follows a corresponding tumor phylogenetic tree, where a multi-regional mutation profile is a matrix in which predictions of somatic mutations at the corresponding tumor regions are listed. Second, we evaluate the lower and upper bounds of specificity and sensitivity of a phylogeny-based somatic variant detection method under several situations. In the evaluation, we conduct patient-wise variant detection from a noisy multi-regional mutation profile matrix for some genomic positions by utilizing the phylogenetic property; we assume that the phylogenetic information can be extracted from another mutation profile matrix that contains accurate candidates at different genomic positions from the noisy ones. From the evaluation, we find that higher sensitivity is not guaranteed in the phylogeny-based variant detection, but higher specificity is guaranteed for several cases. These findings indicate the tumor phylogeny gives more merit for variant detection based on erroneous long-read sequencers (e.g. Oxford nanopore sequencers) than that based on accurate short-read sequencers (e.g., Illumina sequencer).

1 Introduction

Somatic mutation is one of the most important factors for cancer evolution. Acquired somatic mutations, together with individual germline variations, have a large effect on carcinogenesis. By obtaining the accurate profiles of genomic alterations for each patient, we can estimate the cause of cancer and search for optimal therapies for each patient.

© Springer Nature Switzerland AG 2020
G. Bebis et al. (Eds.): ISMCO 2020, LNBI 12508, pp. 87–101, 2020.
https://doi.org/10.1007/978-3-030-64511-3_9

Hence, variant detection from sequence data sets is a fundamental analysis both in cancer therapy and research. There exist two types of variant detection methods: single-tumor-based mutation call and multi-regional one.

In single-tumor-based mutation call, we use one tumor and a matched normal sequence data set for variant detection. For this type of method, a large number of studies [1–8] have been conducted to improve the performance of single-tumor-based mutation call and the performance had been improved annually by modeling properties of raw sequence data sets in more sophisticated manners.

In multi-regional mutation call, we use multiple tumor and a matched normal sequence data sets for variant detection. By leveraging multiple tumor sequence data sets, we can additionally use two properties for mutation calling: sharing property of mutations and the property of tumor phylogeny. multiSNV [9] and MultiMuC [10] are based on the sharing property of mutations. SNV-PPILP [11], Treeomics [12], and MuClone [13] are based on the property of tumor phylogeny.

Although several methods that utilize the property of tumor phylogeny demonstrated improved detection performances, it is ambiguous whether the property always effective for variant detection. A recent study [14] reported SNV-PPILP and MuClone showed poor performances in some situations. These contradictory results require further studies on the property of tumor phylogeny for variant detection; we need to know why and how tumor phylogeny works well for variant detection.

In this paper, we study the performance of a variant detection method that uses the property of tumor phylogeny. First, we give a proof of the property of tumor phylogenetic tree, i.e., the total patterns of mutations are restricted if a multi-regional mutation profile has a corresponding tumor phylogenetic tree. Second, we evaluate the phylogeny-based mutation calling methods under simplified settings.

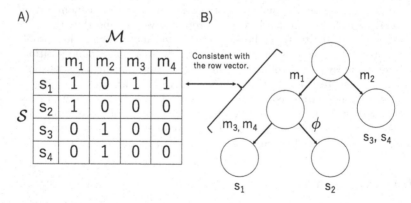

Fig. 1. (A) shows a mutation profile as an example. (B) shows the phylogenetic tree for the mutation profile of (A).

2   Methods

In this section, we first show the property of the tumor phylogenetic tree on the total patterns of mutations after reviewing the definition of tumor phylogenetic tree [15]. We should note that the property on the total patterns of mutations has not been proved by existing phylogeny-based mutation calling methods of SNV-PPILP, Treeomics, and MuClone. Second, we introduce the assumptions for the patient-wise variant detection and evaluate the expected specificity and sensitivity.

2.1   Tumor Phylogenetic Tree

Definition of Tumor Phylogenetic Tree. Here, we review the definition of phylogenetic trees and related topics [15] under infinite sites assumption [16], i.e., each mutation corresponds to only one edge of the tumor phylogenetic tree. For different assumptions, e.g., finite sites assumption and Dollo parsimony, see [17–19]. Let $T \in \{0,1\}^{c \times k}$ be a binary matrix for a mutation profile, where $c \in \mathbb{N}_{>0}$ represents the number of cell types and $k \in \mathbb{N}_{>0}$ represents the number of mutations. A phylogenetic tree is defined for the mutation profile T as follows.

Definition 1. *A phylogenetic tree $\mathcal{T} = (V, E)$ for $T \in \{0,1\}^{c \times k}$ is a rooted tree that satisfies the following condition.*

$$\exists f : \mathcal{S} \to F_{\mathcal{T}}, \exists g : \mathcal{M} \to E, \forall v \in F_{\mathcal{T}}, \forall s \in f^{-1}(\{v\}),$$
$$g^{-1}(P_{\mathcal{T}}(v)) = \{m \in \mathcal{M} | s \text{ has mutation } m\},$$

where

$$V : A \text{ set of vertices of } \mathcal{T}, \quad E : A \text{ set of edges of } \mathcal{T},$$
$$\mathcal{S} : A \text{ set of indices for samples in } T, \quad \mathcal{M} : A \text{ set of indices for mutations in } T,$$
$$F_{\mathcal{T}} : A \text{ set of leaves of } \mathcal{T},$$
$$P_{\mathcal{T}}(v) := \{e \in E | e \text{ is included in the path from the root of } \mathcal{T} \text{ to } v\}.$$

The condition above requires the consistency between the original matrix and the corresponding tumor phylogenetic tree. Figure 1 shows an example of tumor phylogenetic tree for a mutation profile. To determine whether or not the mutation profile has a corresponding tumor phylogenetic tree, we can rely on Lemma 1.

Lemma 1.

$$T \in \{0,1\}^{c \times k} \text{ has a phylogenetic tree}$$
$$\Leftrightarrow \forall i, j \in \{1, \cdots, k\} \text{ s.t. the condition of (i) or (ii) or (iii) is satisfied,}$$
$$\text{(i)} O_i \cap O_j = \phi, \quad \text{(ii)} O_i \subseteq O_j, \quad \text{(iii)} O_i \supseteq O_j,$$

where $O_k := \{s \in \mathcal{S} | s \text{ has mutation } k\}$.

Properties of Tumor Phylogenetic Tree. In the actual multi-regional mutation call from bulk sequence data sets, the tumor phylogenetic tree cannot be
observed directly because multiple types of tumor cells are usually mixed in one
tumor region. To represent the multi-regional mutation calling data set, we con-
sider the variant allele frequency matrix of $Y \in \mathbb{R}_{\geq 0}^{n \times k}$, where n is the number of
sampled tumor regions, k is the number of mutation candidates, and $Y_{i,j}$ rep-
resents the variant allele frequency of the j-th mutation candidate at the i-th
tumor region. Here, we show that any tumor phylogenetic tree restricts the pattern of columns in Y. First, we can assume that the following tumor phylogenetic
tree exists.

Lemma 2.

> If $T \in \{0,1\}^{c \times k}$ has a phylogenetic tree, then
>
> $\exists \mathcal{T} = (V, E)$ s.t. $\mathcal{T}$ satisfies the following conditions,
>
> a) $\mathcal{T}$ is a phylogenetic tree for T,
>
> b) $|F_{\mathcal{T}}| \leq c$,
>
> c) The root node has only one outgoing edge,
>
> d) Any node except for the root has zero or two outgoing edges.

Proof. See appendix section (Sect. A). $\square$

Second, based on the Lemma 2, we can show the following Theorem 1 and
Corollary 1 that lead to the restriction on the pattern of columns in Y.

Theorem 1 (Patterns of Column Vectors in Phylogenetic Matrix).

> If $T \in \{0,1\}^{c \times k}$ has a phylogenetic tree, then $|\{t_i | i = 1, \cdots, k\}| \leq 2c - 1$,
> where t_i is the i-th column vector of T.

Proof. From the definition of the phylogenetic tree under the infinite sites
assumption, one mutation corresponds to only one edge of the tumor phylogenetic tree. Therefore, any mutation in T corresponds to an edge in $\mathcal{T}$. If one
mutation corresponds to an edge in $\mathcal{T}$, we find one pattern from column vectors
in T. If no mutation corresponds, we find no patterns from the column vectors.
Therefore, $|\{t_i | i = 1, \cdots k\}| \leq |E|$. Because of Lemma 2, we can assume that the
root of $\mathcal{T}$ is connected to the root of a full binary tree at which the number of
leaves is $\leq c$. From this, $|E| \leq 2c - 1$. $\square$

Corollary 1 (Observable Cell Types and Column Vectors). *Let $T \in$
$\{0,1\}^{c \times k}$ have a phylogenetic tree, and $U \in \mathbb{R}_{\geq 0}^{n \times c}, U_1 \in \mathbb{R}_{\geq 0}^{n \times c_1}$ have a non-
negative simplex for every row vector, where $U = [U_1 \ O]$. Then, the number of
unique vectors of $(y_1, \cdots y_k)$ in $Y := \frac{1}{2} U T$ satisfies $|\{y_i | i = 1, \cdots k\}| \leq 2c_1 - 1$.*

Proof. We split T by row: $T = \begin{bmatrix} T_1 \\ T_2 \end{bmatrix}$, where $T_1 \in \{0,1\}^{c_1 \times k}, T_2 \in \{0,1\}^{(c-c_1) \times k}$. Then, Y is as follows.

$$Y = \frac{1}{2}UT = \frac{1}{2}U_1 T_1 = \frac{1}{2}(U_1 \tilde{t}_1 \cdots U_1 \tilde{t}_k),$$

where $\tilde{t}_1, \cdots, \tilde{t}_k$ is the column vectors in T_1. Because the equivalent conditions listed in Lemma 1 hold true for the subset of rows, both T_1 and T_2 have a phylogenetic tree. From Theorem 1, the total patterns of column vectors in T_1 is limited by $2c_1 - 1$. Then, $|\{y_i | i = 1, \cdots k\}| \leq 2c_1 - 1$. $\square$

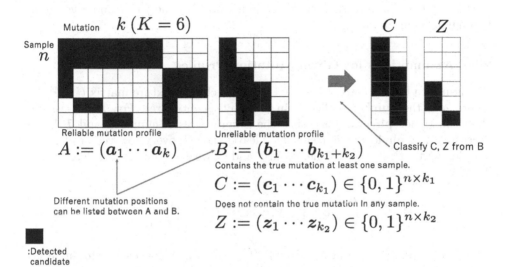

Fig. 2. A graphical summary of the problem settings. In this problem setting, we have a reliable mutation profile A and an unreliable mutation profile B. Within column vectors in B, two types of column vectors exist: column vectors with at least one true mutations (those in C) and those without any mutations (those in Z). The purpose in this problem setting is to label each column vector of B as a column from C or that from Z.

2.2   Assumptions for the Patient-Wise Variant Detection

Given Mutation Profiles. We assume that two types of observed mutation profiles from a patient are given as shown in Fig. 2. The first mutation profile is reliable, e.g., a mutation profile estimated by a short-read sequencer in multi-regional tumors with high depth. We express the first profile as $A \in \{0,1\}^{n \times k}$, where n is the number of sequenced samples and k is the number of mutations.

$A_{n',k'} = 1$ means that the mutation candidate exists at the k'-th genomic position in the n'-th sample. For simplicity, we represent the i-th column vector as a_i, and its j-th element as $a_{i,j}$, and $A = (a_1 \cdots a_k)$.

The second profile is unreliable, e.g., a mutation profile estimated by a long-read sequencer, and this profile contains erroneous positions at which no tumor regions have the true mutation. We describe the second profile as $B \in \{0,1\}^{n \times (k_1+k_2)}$, where k_1 is the number of non-erroneous positions and k_2 is the number of erroneous positions. We make a mutation profile $C \in \{0,1\}^{n \times k_1}$ by collecting only non-erroneous positions from B and make error profile $Z \in \{0,1\}^{n \times k_2}$ by collecting erroneous positions. We represent the j-th column vector as b_j, c_j, z_j, and $B = (b_1 \cdots b_{k_1+k_2})$, $C = (c_1 \cdots c_{k_1})$, $Z = (z_1 \cdots z_{k_2})$. The purpose of the patient-wise variant detection is to label the mutations of B that cannot be listed in A. That is, we judge whether the j-th column vector b_j belongs to C or Z.

2.3   Assumptions for Given Mutation Profiles

We assume a binary matrix $T \in \{0,1\}^{c \times k}$ and clonal mixture matrix $U \in \mathbb{R}_{\geq 0}^{n \times c}$, where c is the number of leaves in the phylogenetic tree. For T, we assume that the tumor phylogeny satisfies the infinite sites assumption and T has a corresponding phylogenetic tree with c leaves. For U, every row vector of U is a simplex vector that sums to one. From T and U, we assume that A is determined. That is,

$$A_{i,j} = h\left(\left(\frac{1}{2}UT\right)_{i,j}\right), \text{ where } h(x) = \begin{cases} 0 & (x = 0) \\ 1 & (x > 0) \end{cases}.$$

For $C \in \{0,1\}^{n \times k_1}$ and $Z \in \{0,1\}^{n \times k_2}$, we assume that each column vector is independently generated by the stochastic models below.

$$I_j \sim \text{DiscreteUnif}(\cdot|1,k),$$
$$\xi_{j,i} \sim \text{Ber}(\cdot|f_1),$$
$$c_{j,i} = \min(a_{I_j,i} + \xi_{j,i}, 1),$$
$$z_{l,i} \sim \text{Ber}(\cdot|f_2),$$

where $\text{DiscreteUnif}(\cdot|a,b)$ represents the discrete uniform distribution with the range from $a \in \mathbb{Z}$ up to $b \in \mathbb{Z}$, $\text{Ber}(\cdot|f)$ is the Bernoulli distribution with frequency of f, $0 < f_1 < 1$, $0 < f_2 < 1$, $j \in \{1, \cdots, k_1\}$, $l \in \{1, \cdots, k_2\}$, and $i \in \{1, \cdots, n\}$. From the above stochastic models, we can see that each column vector c_j has an original template vector a_{I_j} with additive noise ξ_j as shown in Fig. 3, and that each column vector z_j does not have the original template vector.

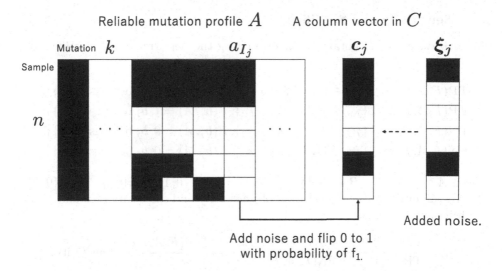

Fig. 3. The assumed generative model of each column vector in C. Letting c_j be the j-th column vector in C, c_j has the original column vector a_{I_j} in A. By adding a noise ξ_j to a_{I_j}, c_j is obtained.

2.4 Labeling Methods

We set a phylogeny-based labeling function $L : \{0,1\}^n \times \{0,1\}^{n \times k} \rightarrow \{0,1\}$ and a sharing-property-based function $R_r : \{0,1\}^n \times \{0,1\}^{n \times k} \rightarrow \{0,1\}$ as follows,

$$
L(\boldsymbol{b}, A) = \begin{cases} 1 & (\exists j \in \{1, \cdots, k\} \ s.t. \ \boldsymbol{b} = \boldsymbol{a}_j) \\ 0 & (\text{Otherwise}) \end{cases}, \tag{1}
$$

$$
R_r(\boldsymbol{b}, A) = \begin{cases} 1 & (\sum_{i=1}^{n} b_i \geq r) \\ 0 & (\text{Otherwise}) \end{cases}. \tag{2}
$$

As we can see from the definition, L sets the label by using A, while R_r sets the label by ignoring A and only use $\boldsymbol{b}$. In other words, L leverages the limited patterns of column vectors of A and R_r leverages the number of detected candidates. Figure 4 shows a graphical summary of labeling methods of L and R_r.

2.5 Sensitivity and Specificity

We introduce several notations for evaluating the performance of classification as follows.

$$\text{TP}(F, A, B) := |\{j | j \in \{1, \cdots, k_1 + k_2\}, F(\boldsymbol{b}_j, A) = 1, \boldsymbol{b}_j \text{ belongs to } C\}|,$$
$$\text{FP}(F, A, B) := |\{j | j \in \{1, \cdots, k_1 + k_2\}, F(\boldsymbol{b}_j, A) = 1, \boldsymbol{b}_j \text{ belongs to } Z\}|,$$
$$\text{TN}(F, A, B) := |\{j | j \in \{1, \cdots, k_1 + k_2\}, F(\boldsymbol{b}_j, A) = 0, \boldsymbol{b}_j \text{ belongs to } Z\}|,$$
$$\text{FN}(F, A, B) := |\{j | j \in \{1, \cdots, k_1 + k_2\}, F(\boldsymbol{b}_j, A) = 0, \boldsymbol{b}_j \text{ belongs to } C\}|,$$

where $A \in \{0,1\}^{n \times k}$, $B \in \{0,1\}^{n \times (k_1 + k_2)}$, and $F : \{0,1\}^n \times \{0,1\}^{n \times k} \to \{0,1\}$.

We will evaluate the expectation of sensitivity and specificity of L and R_r in the following sections.

$$\mathbb{E}_{B|A}\left[\frac{\text{TP}(F, A, B)}{\text{TP}(F, A, B) + \text{FN}(F, A, B)}\right] = \frac{\mathbb{E}_{B|A}[\text{TP}(F, A, B)]}{k_1} \text{ (Sensitivity)},$$

$$\mathbb{E}_{B|A}\left[\frac{\text{TN}(F, A, B)}{\text{FP}(F, A, B) + \text{TN}(F, A, B)}\right] = \frac{\mathbb{E}_{B|A}[\text{TN}(F, A, B)]}{k_2} \text{ (Specificity)},$$

where expectation $\mathbb{E}_{B|A}$ is taken with respect to all the generated unreliable mutation profile B with given reliable mutation profile A.

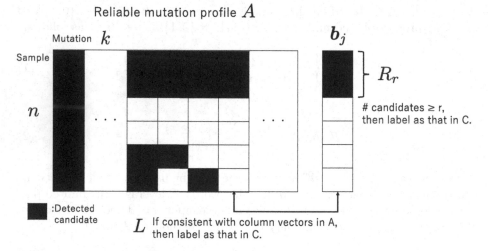

Fig. 4. A graphical summary for L and R_r. L checks the existence of a consistent column vector in A. R_r only checks the number of detected candidate in a given column vector from B.

3   Performance Evaluation

3.1   Performance Evaluation Summary of L, R_r

Under the assumption described before, the expected value of specificity and sensitivity for L, R_r can be summarized as follows.

$$\frac{\mathbb{E}_{B|A}[\mathrm{TN}(R_r, A, B)]}{k_2} = 1 - \sum_{x=r}^{n} {}_nC_x \, (1-f_2)^{n-x} f_2^x, \tag{3}$$

$$\frac{\mathbb{E}_{B|A}[\mathrm{TP}(R_r, A, B)]}{k_1} = \sum_{q=r}^{n} \sum_{x=1}^{q} w_x \, {}_{n-x}C_{q-x} \, f_1^{q-x}(1-f_1)^{n-q}, \tag{4}$$

$$(1 - K\overline{f_2}^n) \le \frac{\mathbb{E}_{B|A}[\mathrm{TN}(L, A, B)]}{k_2} \le (1 - K\underline{f_2}^n), \tag{5}$$

$$G_n(\boldsymbol{w}, (1-f_1)) \le \frac{\mathbb{E}_{B|A}[\mathrm{TP}(L, A, B)]}{k_1} \le KG_n(\boldsymbol{w}, \overline{f_1}), \tag{6}$$

where

K : The total patterns of columns in A,

$$\boldsymbol{w} := (w_1, \cdots, w_n), \ \ w_i := \Pr\left(\boldsymbol{c}_j \ s.t. \sum_{n'=1}^{n} a_{I_j,n'} = i\right),$$

$$\overline{f_1} := \max(f_1, 1-f_1), \ \ \overline{f_2} := \max(f_2, 1-f_2), \underline{f_2} := \min(f_2, 1-f_2),$$

$$G_n(\boldsymbol{x}, f) := \sum_{i=1}^{n} x_i f^{(n-i)}, \ \ {}_nC_x := \frac{n!}{x!(n-x)!}.$$

4   Results

Here, we would like to evaluate the lower bounds in Eqs. (6) and (5) at several settings of f_1, f_2, $\boldsymbol{w}$, n, K. For $\boldsymbol{w}$, we sample $\boldsymbol{w}$ from Dirichlet$(\cdot|(\alpha, \cdots, \alpha))$, where $\alpha = 1.0$, and take the average of the lower bounds. The following procedure is used for evaluation.

1. Conduct the following procedures 100 times and take the average of the lower bounds for expected specificity and sensitivity for each f_1, f_2, n, K.
   1-a) Sample $\boldsymbol{w} \sim$ Dirichlet$(\cdot|(\alpha, \cdots, \alpha))$, where $\alpha = 1.0$.
   1-b) Evaluate the lower bound of $\frac{\mathbb{E}_{B|A}[\mathrm{TP}(L,A,B)]}{k_1}$ and $\frac{\mathbb{E}_{B|A}[\mathrm{TN}(L,A,B)]}{k_2}$.
   1-c) If the lower bound > 1.0, substitute 1.0 for the bound.
   1-d) If the lower bound < 0.0, substitute 0.0 for the bound.

The performance evaluation results of L are shown in Fig. 5. For the specificity of L, when f_2 is around 0.5, detection specificity is high and the evaluated bounds are meaningless when f_2 is close to 0 or 1. For the sensitivity of L, detection sensitivity is from 5% to 40%. From this, we can detect a somatic mutation in a patient with high specificity and moderate (but not ignorable) sensitivity. Similarly, we also evaluate the performence of R_r in the appendix at Fig. 7.

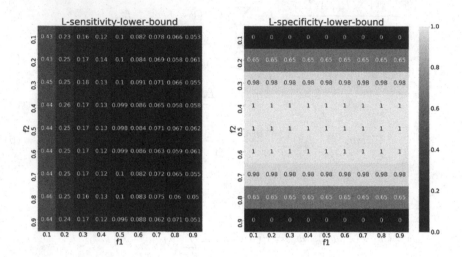

Fig. 5. Lower bounds of specificity and sensitivity of L at $n = 20$, $K = 30$.

5   Conclusion

In this paper, we studied the performance of patient-wise somatic variant detection in a phylogeny-based manner. As a result, tumor phylogeny is effective to achieve higher detection specificity in several cases, and the phylogeny is not effective to achieve higher detection sensitivity. This result indicates tumor phylogeny might be more useful for variant detection from erroneous long-read sequence data sets if corresponding accurate short-read sequence data sets exist.

Acknowledgments. This work has been supported by the Grant-in-Aid for JSPS Research Fellow (17J08884) and MEXT/JSPS KAKENHI Grant (15H05912, hp180198, hp170227, 18H03329, hp190158).

A   Appendix

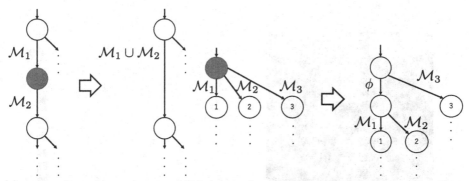

(a) A procedure of removing a node having only one outgoing edge.

(b) A procedure of removing a node having more than two outgoing edges.

Fig. 6. Procedures for removing a node having only one outgoing edge and removing a node having more than two outgoing edges.

Proof. (proof of Lemma Lemma 2) T has a phylogenetic tree, hence we can choose a phylogenetic tree $\mathcal{T}$. We can assume $|F_{\mathcal{T}}| \leq c$ by removing leaves in $\mathcal{T}$ if no cell corresponds to the leaf in $f : R \to F_{\mathcal{T}}$. We can also assume that the root node has only one outgoing edge by adding a new root node and connect the novel root and the previous root node.

For the last condition, we remove the following two types of internal nodes: i) the internal node having only one outgoing edge, and ii) the internal node having more than 2 outgoing edges. It is sufficient to show the operation to remove nodes that satisfy i) or ii) from T while keeping conditions of a)-c).

For i), just remove the nodes as in Fig. 6a. We can easily check a)-c) still holds true after this operation. For ii) just remove the node as in Fig. 6b. If the number of outgoing edges is more than three, apply this operation recursively. We can also check that a)-c) still hold true after these operations. Because the operations pictured in Fig. 6a and 6b just decrease the number of i) and ii) nodes, we can totally remove i) and ii) nodes. □

A.1   Performance Evaluation of R_r

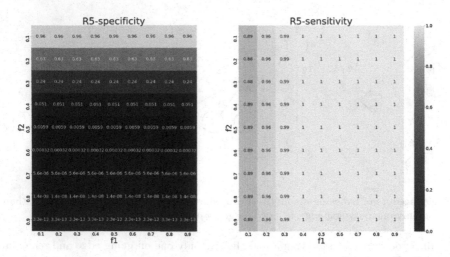

Fig. 7. Specificity and sensitivity of R_r at $r = 5$, $n = 20$, $K = 30$.

A.2   Detailed Procedures for Performance Evaluation

Evaluation of $\mathbb{E}_{B|A}[\mathrm{TN}(L, A, B)]/k_2$. We evaluate the upper bound and lower bound for $\mathbb{E}_{B|A}[\mathrm{FP}(L, A, B)]$. Letting K be the total patterns of columns in A, the lower bound can be derived as follows.

$$
\mathbb{E}_{B|A}[\mathrm{FP}(L, A, B)] = k_2 \sum_{j=1}^{K} \left(\prod_{i=1}^{n} f_2^{a_{I_j,i}} (1 - f_2)^{1 - a_{I_j,i}} \right)
$$

$$
\geq k_2 K \min_{j \in \{1, \cdots, K\}} \left(\prod_{i=1}^{n} f_2^{a_{I_j,i}} (1 - f_2)^{1 - a_{I_j,i}} \right)
$$

$$
\geq k_2 K \min(f_2, 1 - f_2)^n = k_2 K \underline{f_2}^n,
$$

where $\underline{f_2} := \min(f_2, 1 - f_2)$. The upper bound can also be derived as follows.

$$
\mathbb{E}_{B|A}[\mathrm{FP}(L, A, B)] = k_2 \sum_{j=1}^{K} \left(\prod_{i=1}^{n} f_2^{a_{I_j,i}} (1 - f_2)^{1 - a_{I_j,i}} \right)
$$

$$
\leq k_2 K \max_{j \in \{1, \cdots, K\}} \left(\prod_{i=1}^{n} f_2^{a_{I_j,i}} (1 - f_2)^{1 - a_{I_j,i}} \right)
$$

$$
\leq k_2 K \max(f_2, 1 - f_2)^n = k_2 K \overline{f_2}^n,
$$

where $\overline{f_2} := \max(f_2, 1 - f_2)$. From this, we can evaluate $\mathbb{E}_{B|A}[\mathrm{TN}(L, A, B)]$ as follows.

$$(1 - K\overline{f_2}^n) \leq \frac{\mathbb{E}_{B|A}[\mathrm{TN}(L, A, B)]}{k_2} \leq (1 - K\underline{f_2}^n). \tag{7}$$

Evaluation of $\mathbb{E}_{B|A}[\mathrm{TP}(L, A, B)]/k_1$. From the linearity of the expectation, the expected number of true positives can be written as follows.

$$\mathbb{E}_{B|A}[\mathrm{TP}(L, A, B)] = \mathbb{E}_{B|A}\left[\sum_{j=1}^{k_1} L(\boldsymbol{c}_j, A)\right] = \sum_{j=1}^{k_1} \Pr(L(\boldsymbol{c}_j, A) = 1).$$

The lower bound for $\Pr(L(\boldsymbol{c}_j, A) = 1)$ is as follows.

$$\Pr(L(\boldsymbol{c}_j, A) = 1)$$

$$= \sum_{i=1}^{n} \Pr\left(L(\boldsymbol{c}_j, A) = 1, \boldsymbol{c}_j \ s.t. \ \sum_{n'=1}^{n} a_{I_j, n'} = i\right)$$

$$= \sum_{i=1}^{n} \Pr\left(L(\boldsymbol{c}_j, A) = 1 \middle| \boldsymbol{c}_j \ s.t. \ \sum_{n'=1}^{n} a_{I_j, n'} = i\right) \Pr\left(\boldsymbol{c}_j \ s.t. \ \sum_{n'=1}^{n} a_{I_j, n'} = i\right)$$

$$= \sum_{i=1}^{n} w_i \Pr\left(L(\boldsymbol{c}_j, A) = 1 \middle| \boldsymbol{c}_j \ s.t. \ \sum_{n'=1}^{n} a_{I_j, n'} = i\right)$$

$$\geq \sum_{i=1}^{n} w_i \Pr\left(\boldsymbol{a}_{I_j} = \boldsymbol{c}_j \middle| \boldsymbol{c}_j \ s.t. \ \sum_{n'=1}^{n} a_{I_j, n'} = i\right) = \sum_{i=1}^{n} w_i (1 - f_1)^{(n-i)},$$

$$\because) \ \boldsymbol{a}_{I_j} = \boldsymbol{c}_j \Rightarrow L(\boldsymbol{c}_j, A) = 1,$$

where $w_i := \Pr\left(\boldsymbol{c}_j \ s.t. \ \sum_{n'=1}^{n} a_{I_j, n'} = i\right)$. From this,

$$\mathbb{E}_{B|A}[\mathrm{TP}(L, A, B)] \geq k_1 \sum_{i=1}^{n} w_i (1 - f_1)^{(n-i)}.$$

For obtaining the upper bound of $\mathbb{E}_{B|A}[\mathrm{TP}(L, A, B)]$, we focus on two things as shown in Fig. 8. First, the number of column vectors in A that each $\boldsymbol{c}_j$ can correspond is at most K. Second, the probability for each $\boldsymbol{c}_j$ corresponding to one column vector is at most $\overline{f_1}^{n-i}$, where $\overline{f_1} := \max(f_1, 1 - f_1)$, and $i = \sum_{n'=1}^{n} a_{I_j, n'}$. From this, we can obtain the upper bound for the conditional probability as follows.

$$\Pr\left(L(\boldsymbol{b}_j, A) = 1 \middle| \boldsymbol{c}_j \ s.t. \ \sum_{n'=1}^{n} a_{I_j, n'} = i\right) \leq K\overline{f_1}^{(n-i)}.$$

Then, the upper bound of $\mathbb{E}_{B|A}[\mathrm{TP}(L, A, B)]$ is as follows.

$$\mathbb{E}_{B|A}[\mathrm{TP}(L, A, B)] \leq k_1 \sum_{i=1}^{n} w_i K\overline{f_1}^{(n-i)} = k_1 K \sum_{i=1}^{n} w_i \overline{f_1}^{(n-i)}.$$

Therefore,

$$G_n(\boldsymbol{w}, (1 - f_1)) \leq \frac{\mathbb{E}_{B|A}[\mathrm{TP}(L, A, B)]}{k_1} \leq K G_n(\boldsymbol{w}, \overline{f_1}),$$

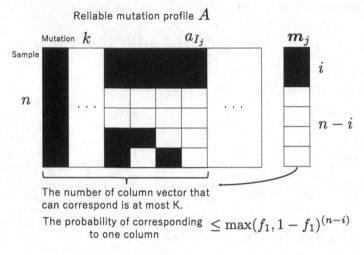

Reliable mutation profile A

Fig. 8. The key idea for obtaining the upper bound of $\mathbb{E}_{B|A}[\mathrm{TP}(L, A, B)]$.

Performance Evaluation of R_r. We can evaluate the specificity and sensitivity for R_r.

$$\frac{\mathbb{E}_{B|A}[\mathrm{TP}(R_r, A, B)]}{k_1} = (k_1)^{-1} \mathbb{E}_{B|A} \left[\sum_{j=1}^{k_1} \sum_{q=r}^{n} \mathbb{I}_{\left\{ \sum_{n'=1}^{n} c_{j,n'} = q \right\}} \right]$$

$$= (k_1)^{-1} \sum_{j=1}^{k_1} \sum_{q=r}^{n} \Pr \left(\sum_{n'=1}^{n} c_{j,n'} = q \right)$$

$$= (k_1)^{-1} \sum_{j=1}^{k_1} \sum_{q=r}^{n} \sum_{x=1}^{q} \Pr \left(\sum_{n'=1}^{n} b_{j,n'} = q, c_j \ s.t. \sum_{n''=1}^{n} a_{I_j, n''} = x \right)$$

$$= (k_1)^{-1} \sum_{j=1}^{k_1} \sum_{q=r}^{n} \sum_{x=1}^{q} w_x \ {}_{n-x}C_{q-x} \ f_1^{q-x} (1 - f_1)^{n-q}$$

$$= \sum_{q=r}^{n} \sum_{x=1}^{q} w_x \ {}_{n-x}C_{q-x} \ f_1^{q-x} (1 - f_1)^{n-q},$$

$$\frac{\mathbb{E}_{B|A}[\mathrm{TN}(R_r, A, B)]}{k_2} = \sum_{x=0}^{r-1} {}_nC_x \ (1 - f_2)^{n-x} \ f_2^x = 1 - \sum_{x=r}^{n} {}_nC_x \ (1 - f_2)^{n-x} \ f_2^x.$$

References

1. Koboldt, D.C., et al.: VarScan 2: somatic mutation and copy number alteration discovery in cancer by exome sequencing. Genome Res. **22**(3), 568–576 (2012)
2. Saunders, C.T., et al.: Strelka: accurate somatic small-variant calling from sequenced tumor-normal sample pairs. Bioinformatics **28**(14), 1811–1817 (2012)
3. Cibulskis, K., et al.: Sensitive detection of somatic point mutations in impure and heterogeneous cancer samples. Nat. Biotechnol. **31**(3), 213–219 (2013)
4. Shiraishi, Y., et al.: An empirical Bayesian framework for somatic mutation detection from cancer genome sequencing data. Nucleic Acids Res. **41**(7), e89 (2013)
5. Usuyama, N., et al.: HapMuC: somatic mutation calling using heterozygous germ line variants near candidate mutations. Bioinformatics **30**(23), 3302–3309 (2014)
6. Kim, S., et al.: Strelka2: fast and accurate calling of germline and somatic variants. Nat. Methods **15**(8), 591–594 (2018)
7. Moriyama, T., et al.: A Bayesian model integration for mutation calling through data partitioning. Bioinformatics **35**(21), 4247–4254 (2019)
8. Sahraeian, S.M.E., et al.: Deep convolutional neural networks for accurate somatic mutation detection. Nat. Commun. **10**(1), 1041 (2019)
9. Josephidou, M., et al.: multiSNV: a probabilistic approach for improving detection of somatic point mutations from multiple related tumour samples. Nucleic Acids Res. **43**(9), e61 (2015)
10. Moriyama, T., et al.: Accurate and flexible bayesian mutation call from multi-regional tumor samples. In: Mathematical and Computational Oncology, pp. 47–61. Springer, Cham (2019)
11. van Rens, K.E., et al.: SNV-PPILP: refined SNV calling for tumor data using perfect phylogenies and ILP. Bioinformatics **31**(7), 1133–1135 (2015)
12. Reiter, J.G., et al.: Reconstructing metastatic seeding patterns of human cancers. Nat. Commun. **8**, 14114 (2017)
13. Dorri, F., et al.: Somatic mutation detection and classification through probabilistic integration of clonal population information. Commun. Biol. **2**(1), 44 (2019)
14. Detering, H., et al.: Accuracy of somatic variant detection in multiregional tumor sequencing data. bioRxiv 655605 (2019)
15. Gusfield, D.: Efficient algorithms for inferring evolutionary trees. Networks **21**(1), 19–28 (1991)
16. Kimura, M.: The number of heterozygous nucleotide sites maintained in a finite population due to steady flux of mutations. Genetics **61**(4), 126 (1969)
17. Zafar, H., et al.: SiFit: inferring tumor trees from single-cell sequencing data under finite-sites models. Genome Biol. **18**(1), 178 (2017)
18. Zafar, H., et al.: SiCloneFit: bayesian inference of population structure, genotype, and phylogeny of tumor clones from single-cell genome sequencing data. Genome Research (2019)
19. El-Kebir, M.: SPhyR: tumor phylogeny estimation from single-cell sequencing data under loss and error. Bioinformatics **34**(17), i671–i679 (2018)

Detecting Subclones from Spatially Resolved RNA-Seq Data

Phillip B. Nicol[✉]

Harvard University, Cambridge, MA 02138, USA
phillipnicol@college.harvard.edu

Abstract. Recently developed technologies allow us to view the transcriptome at high resolution while preserving the spatial location of samples. These advances are particularly relevant to cancer research, since clonal theory predicts that nearby cells are likely to belong to the same expanding subclone. Using this evolutionary hypothesis, we develop a statistical procedure which uses a test of local spatial association along with a graph-based approach to infer subclones from spatially resolved RNA-seq data. Our method is robust, scalable, and can be applied to data from any of the existing spatial transcriptomics technologies. On data from spatially resolved RNA-seq of breast cancer tissue, our method infers seven distinct subclones and identifies potential driver genes.

Keywords: Spatial transcriptomics · Cancer evolution · Spatial statistics

1 Introduction

Tumors grow through a series of clonal expansions [7]. When individual tumor cells acquire advantageous mutations, subclonal expansions begin as these cells proliferate and outcompete surrounding tumor cells. The result of this process is a highly heterogeneous tumor population which is comprised of many subclones. Locally, however, the tumor population will be homogeneous since nearby cells are likely to belong to the same expanding subclone. Statistical methods which can leverage our evolutionary knowledge about the spatial structure of the tumor population may be able to make more accurate predictions, which could in turn lead to improved treatments and therapeutics.

Recently developed technologies can perform RNA sequencing while preserving the spatial relationship between cells or samples. seqFISH can obtain subcellular resolution of a targeted subset of the transcriptome [12], although the cost and complexity may prohibit widespread use [3]. Spatial Transcriptomics (ST) can sequence the entire transcriptome at a collection of regularly spaced sites which may contain several hundred cells [14]. In 2019, Rodriques and colleagues introduced Slide-seq, which is able to sequence the entire transcriptome at the single cell level [11].

© Springer Nature Switzerland AG 2020
G. Bebis et al. (Eds.): ISMCO 2020, LNBI 12508, pp. 102–107, 2020.
https://doi.org/10.1007/978-3-030-64511-3_10

Several statistical methods have been developed to identify genes which exhibit spatially variable patterns of expression [5,15]. However, these methods report only a global measure of spatial association, which is insufficient for determining the location and composition of possible tumor subclones. Elyanow and colleagues recently introduced a method based on hidden Markov random fields (HMRFs) to infer the location of subclones from ST data at the resolution of copy number variation [6]. A similar method was used by [16] to find spatially associated cell populations in seqFISH data.

In this paper, we introduce a new statistical method for inferring subclones from spatially resolved RNA-seq data. Our method uses the local Moran's I statistic to determine locations for which a gene may be expressed as part of a subclone and then uses a graph-based approach to cluster genes with similar patterns of spatial expression. Importantly, our method is flexible and efficient enough to be applied to data from any of the existing spatial RNA-seq technologies. On a ST breast cancer dataset our method infers seven distinct subclones, identifies potential driver genes, and gives hints about the evolutionary relationship between subclones.

2 Methods

In spatial RNA-seq, we observe the expression levels for n genes at m locations $s^1, \ldots, s^m \in \mathbb{R}^2$. We assume that a population of normal tissue has also been measured so that the relative expression of the cancer tissue can be calculated.

If a differentially expressed gene at location s^i characterizes an expanding subclone we would expect nearby sites to have similar expression levels. We use the *local Moran's I* statistic [2] to quantify the spatial autocorrelation around s^i. The statistic is defined as

$$I_i = z_i \sum_{j=1}^{m} w_{ij} z_j \tag{1}$$

where z_i is the z-score for the relative gene expression level at s^i. The weight w_{ij} encodes the spatial dependency between location i and j. We select w_{ij} to be inversely proportional to the squared Euclidean distance between s^i and s^j with $w_{ii} = 0$.

Under the null hypothesis of no spatial association, Anselin proves the asymptotic normality of (1) and calculates the mean and variance [2]. However, the approximation can be poor in practice and the preferred approach is to use a permutation test in which the z-scores are randomly shuffled (such that z_i stays fixed) and the statistic is recomputed. This generates a null distribution for which an approximate p-value can be computed.

Once the above test has been performed, we obtain a collection of sites for which the gene exhibits positive spatial autocorrelation. For gene i, define $x^i \in \{0,1\}^m$ such that the j-th slot of x^i is 1 if the null hypothesis of no positive spatial autocorrelation is rejected at site s^j. Genes with similar clustering patterns can be identified using Fisher's exact test. Suppose $|x^i| = r_i$ and $|x^j| = r_j$

($|\cdot|$ is the ℓ_1 norm). Under the null hypothesis of independence, the probability of k sites in which both x^i and x^j are 1 follows a hypergeometric distribution $\binom{r_i}{k}\binom{m-r_i}{r_j-k}/\binom{m}{r_j}$. Performing this test for all pairs of genes, we obtain an undirected graph $G = (V, E)$ where vertices are genes and $(i, j) \in E$ if the null hypothesis of independence between x^i and x^j is rejected. We expect that a set of genes characterizing a tumor subclone forms a densely connected region of G. To make this precise, we define $C_i \subset V$ to be the maximal clique containing gene i. A *clique* C is a subset of vertices such that $(c, c') \in E$ for all $c, c' \in C$ and a clique is maximal if it cannot be extended to a larger clique. The final step of the method clusters the C_i by converting them to binary n-dimensional vectors. Since these cliques are in one-to-one correspondence with genes, we consider the resulting clusters to define gene sets which characterize distinct subclones. A flowchart summarizing the main steps of the proposed method is shown in Fig. 1.

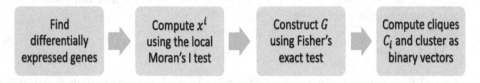

Fig. 1. Flowchart describing the four main steps of the proposed method. For each differentially expressed gene i, the local Moran's I test finds the locations for which the gene expression values are spatially autocorrelated and encodes these locations as the binary vector x^i. The x^i are then compared pairwise using Fisher's exact test, and an edge (i, j) is added to G if x^i and x^j are found to be significantly associated. Finally, we compute C_i (the maximal clique in G containing i) and cluster them as binary vectors.

3   Results

We apply our procedure to the ST breast cancer data provided by [14] which is freely available at www.spatialresearch.org. The dataset contains 4 different layers of tissue and does not provide spatial information about the distances between layers. We thus set $w_{ij} = 0$ whenever sites i and j belong to different layers. Normal tissue was also sequenced, but the dataset does not label cell types. We identify cancer sites by overexpression of *ERBB2* (which codes for *HER2*), a well-known biomarker of breast cancer [13]. Comparing to the provided hematoxylin and eosin (HE) stain, this method of separation appears to be accurate and labels 220 of the 1031 sites as cancer.

We begin by using the Bioconductor package DESEQ2 [9] to find differentially expressed genes in the cancer tissue, removing genes with an adjusted p-value above 0.05. Additionally, we remove genes with non-zero counts in fewer than 20 sites. These two steps reduce the number of genes in consideration from 16391 to 5398. To compute relative expression, we first scale the counts at each site to sum to the median library size over all sites. We then perform a $\log(1 + x)$ transformation and subtract the mean of the control tissue from the cancer sites. We consider a gene to be differentially expressed at s^i if its relative expression exceeds $\log_2(3/2)$. For the local Moran's I test, we use 200 permutations and reject the null hypothesis if the p-value is less than 0.05. From this, we compute the x^j (as described in Methods) for each gene.

There are 278 genes for which $|x^i| \geq 10$, and we restrict our attention to this subset of genes for further analysis. Gene *PRSS23* was found to have significant spatial association at 85 cancer sites, more than any other gene. To construct the graph G, we use the hypergeometric test with an FDR cutoff of 0.05. This yields a graph with 5367 edges. We then compute the maximal clique containing each gene using the CRAN package IGRAPH [4] and further restrict to the genes for which the corresponding clique has size at least 10. We then cluster these cliques and visualize the result using t-SNE [10]. This approach yields 7 distinct clusters, with the largest (cluster 2) containing 60 genes. We consider these clusters to constitute distinct tumor subclones. For plotting, we say that a subclone occupies s^i if at least 10 of the genes in that subclone are differentially expressed at s^i. Figure 2 plots the subclones underneath each of the HE stains.

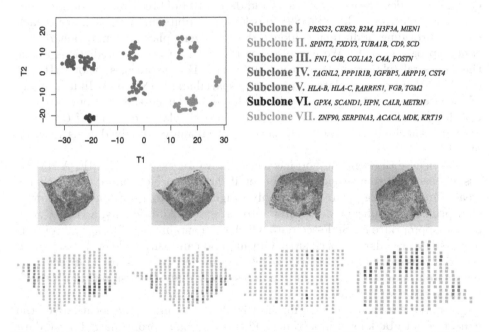

Subclone I. *PRSS23, CERS2, B2M, H3F3A, MIEN1*
Subclone II. *SPINT2, FXDY3, TUBA1B, CD9, SCD*
Subclone III. *FN1, C4B, COL1A2, C4A, POSTN*
Subclone IV. *TAGNL2, PPP1R1B, IGFBP5, ARPP19, CST4*
Subclone V. *HLA-B, HLA-C, RARRES1, FGB, TGM2*
Subclone VI. *GPX4, SCAND1, HPN, CALR, METRN*
Subclone VII. *ZNF90, SERPINA3, ACACA, MDK, KRT19*

Fig. 2. Detecting subclones in breast cancer data. (Top left) t-SNE embedding of cliques containing more than 10 genes. Cliques were converted into binary vectors and clustered by ℓ_2 norm using the CRAN package MERCATOR [1]. (Top right) Five genes from each subclone. (Bottom) HE stains for each of the four layers (taken from [14]) and plots showing the locations of inferred subclones.

To test the robustness of our method, we apply the above procedure to the breast cancer dataset with spatial locations randomly permuted. This shuffling removes all spatial association, so we expect no subclones to exist in this dataset. Since we choose not to correct for multiplicity in the local Moran's I test, there are 23 genes for which $|x^i| \geq 10$. Despite this, the inferred gene graph G has only 10 edges and a maximum clique of size 3, and thus no subclones are inferred. This suggests that the clique-based approach controls for false positives introduced during the local Moran's I test.

We also consider a semi-simulated approach to test the ability of our method to detect artificially created subclones. We select a site s^* and change 20 genes to be overexpressed. To simulate the replication of this cell, we create 50 new sites with the same expression profiles at locations sampled from a normal distribution $\mathcal{N}(s^*, I_{2 \times 2})$. Our method perfectly recovers the artificial subclone by assigning the 50 simulated sites to a new subclone and by identifying the 20 artificially overexpressed genes as markers of this subclone.

4   Discussion

Our method identifies *PRSS23* as a candidate driver of breast cancer progression. A recent study found that *PRSS23* knockdown inhibits tumor growth in stomach cancer, providing additional evidence that this gene plays an important role in cancer progression [8]. We can also use the results to make predictions about the ancestral relationships between subclones. For instance, subclone III and subclone V are closely related on both the spatial and t-SNE plots. In fact, many of the sites are assigned to both subclone III and V, suggesting that one subclone could have evolved from the other. Our approach places *FN1* and *POSTN* in the same subclone which agrees with the original analysis of the dataset (Fig. 4E of [14]).

We believe that our method has several advantages over a HMRF. A HMRF assigns each cell to a subpopulation even if the evidence of spatial association is weak. On the contrary, our method only assigns a gene to a subclone if its expression profile is significantly spatially autocorrelated. Additionally, a HMRF may not be appropriate for Slide-seq data which can contain tens of thousands of cells at possibly irregularly spaced sites. Our method can easily scale to datasets with tens of thousands of samples by parallelizing the permutation test. A direct comparison on simulated data would be necessary to fully understand the tradeoffs between the two approaches.

A major limitation of our method is that it requires the locations of the cancer cells to be known in advance. This is especially problematic for ST data since each site can contain a mixture of different cell types. Additional analysis is therefore needed to determine if any of the inferred subclones in Fig. 2 arose from a failure to control for non-cancer cells. A future improvement to the method could be the use of an explicit evolutionary model to select the spatial weights w_{ij}. An R package implementing the method is available at https://github.com/phillipnicol/SpaceClones3.

References

1. Abrams, Z.B., Coombes, C.E., Li, S., Coombes, K.R.: Mercator: an r package for visualization of distance matrices. bioRxiv (2019). https://doi.org/10.1101/733261
2. Anselin, L.: Local indicators of spatial association-lisa. Geographical Anal. **27**(2), 93–115 (1995). https://doi.org/10.1111/j.1538-4632.1995.tb00338.x
3. Asp, M., Bergenståhle, J., Lundeberg, J.: Spatially resolved transcriptomes-next generation tools for tissue exploration. BioEssays (2020). https://doi.org/10.1002/bies.201900221
4. Csardi, G., Nepusz, T.: The igraph software package for complex network research. Int. J. Complex Syst. **1695** (2006), http://igraph.org
5. Edsgärd, D., Johnsson, P., Sandberg, R.: Identification of spatial expression trends in single-cell gene expression data. Nat. Methods **15**(5), 339–342 (2018). https://doi.org/10.1038/nmeth.4634
6. Elyanow, R., Zeira, R., Land, M., Raphael, B.J.: Starch: copy number and clone inference from spatial transcriptomics data. bioRxiv (2020). https://doi.org/10.1101/2020.07.13.188813
7. Greaves, M., Maley, C.C.: Clonal evolution in cancer. Nature **481**(7381), 306–313 (2012). https://doi.org/10.1038/nature10762
8. Han, B., Yang, Y., Chen, J., He, X., Lv, N., Yan, R.: Prss23 knockdown inhibits gastric tumorigenesis through eif2 signaling. Pharmacol. Res. **142**, 50–57 (2019). https://doi.org/10.1016/j.phrs.2019.02.008
9. Love, M.I., Huber, W., Anders, S.: Moderated estimation of fold change and dispersion for rna-seq data with deseq2. Genome Biol. **15**(12), 550 (2014). https://doi.org/10.1186/s13059-014-0550-8
10. Maaten, L.V.D., Hinton, G.: Visualizing data using t-sne. J. Machine Learn. Res. **9**, 2579–2605 (2008)
11. Rodqiques, S.G., et al.: Slide-seq: a scalable technology for measuring genome-wide expression at high spatial resolution. Science **363**(6434), 1463–1467 (2019). https://doi.org/10.1126/science.aaw1219
12. Shah, S., Lubeck, E., Zhou, W., Cai, L.: In situ transcription profiling of single cells reveals spatial organization of cells in the mouse hippocampus. Neuron **92**(2), 342–357 (2016). https://doi.org/10.1016/j.neuron.2016.10.001
13. Slamon, D.J., Clark, G.M., Wong, S.G., Levin, W.J., Ullrich, A., McGuire, W.L.: Human breast cancer: correlation of relapse and survival with amplification of the her-2/neu oncogene. Science **235**(4785), 177–182 (1987)
14. Ståhl, P.L., et al.: Visualization and analysis of gene expression in tissue sections by spatial transcriptomics. Science **353**(6294), 78–82 (2016). https://doi.org/10.1126/science.aaf2403
15. Sun, S., Zhu, J., Zhou, X.: Statistical analysis of spatial expression patterns for spatially resolved transcriptomic studies. Nature Methods **17**(2), 193–200 (2020). https://doi.org/10.1038/s41592-019-0701-7
16. Zhu, Q., Shah, S., Dries, R., Cai, L., Yuan, G.C.: Identification of spatially associated subpopulations by combining scrna-seq and sequential fluorescence in situ hybridization data. Nature biotechnology (2018)

Novel Driver Synonymous Mutations in the Coding Regions of GCB Lymphoma Patients Improve the Transcription Levels of BCL2

Ofek Shami-Schnitzer[1], Zohar Zafir[2], and Tamir Tuller[2(✉)]

[1] School of Computer Science, Tel Aviv University, 69978 Tel-Aviv, Israel
[2] Department of Biomedical Engineering, the Engineering Faculty, Tel Aviv University, 69978 Tel-Aviv, Israel
tamirtul@tauex.tau.ac.il

Abstract. Synonymous mutations inside the coding region which do not alter the amino acid chain are usually considered to have no effect on the protein. However, in recent years it was shown that they may regulate expression levels via various mechanisms, suggesting that they may also play an important role in tumorigenesis.

In the current study, we suggest a pipeline for detecting cancerous synonymous mutations that affect the cancer fitness via regulation of transcription. We demonstrate our approach by reporting cases where cancerous synonymous mutations inside the coding regions of the gene BCL2 are under selection in germinal center B-cell (GCB) lymphoma patients. We provide various lines of evidence that suggest that these mutations contribute to the cancer cell survival via improving the expression levels of the anti-apoptotic BCL2 gene.

1 Introduction

Mutations in the DNA are considered to be a focal factor causing neoplasm and therefore cancer [1]. Mutations in a protein coding gene could result in a change of the amino acid chain (*i.e.* missense mutations) or be synonymous and only change the nucleotide sequence without altering the sequence of the protein amino acids. While synonymous or 'silent' mutations are often considered to have no effect on the protein sequence, there is evidence suggesting their significant effect on gene function [2], and play a crucial part in development of cancer [3–8]; e.g., Supek *et al.* showed that synonymous mutations frequently contribute to human cancer by altering exonic motifs that regulate splicing in oncogenes; moreover, they might affect regulation signals at the RNA-level as microRNA [3] or at the DNA-level as transcription factors [9].

Up to date, there are more than 700 documented genes found to be associated with cancer development through somatic mutations [10]. The BCL2 family is a group of genes, which among others are responsible for apoptosis, the cell inner death mechanism. One of the major members of this family is the BCL2 (B-Cell lymphoma 2) gene;

G. Bebis et al. (Eds.): ISMCO 2020, LNBI 12508, pp. 108–118, 2020.
https://doi.org/10.1007/978-3-030-64511-3_11

this gene acts as pro-survival gene as it prevents the release of cytochrome C and therefore prevents the activation of caspase 9 that promotes apoptosis. As a key pro-survival gene, malfunction of BCL2 may cause failure in the apoptosis system [11, 12]. Overexpression of the BCL2 gene was found to be connected to large diffuse B-cell lymphoma (DLBCL) [13], the most common type of non-Hodgkin lymphoma. Translocation of BCL2(14;18)(q32;q21) was found as the reason for the overexpression in 20–30% of DLBCL cases [13, 14]. In addition, translocation of the BCL2 gene contributes to the creation of mutations, especially synonymous mutations. Moreover it is apparent that there is selection in these synonymous mutations towards certain mutations [14].

Consequently, we hypothesize that other mechanisms contribute to the overexpression of the BCL2 gene, such as synonymous mutations. However, no research has investigated the influence in DLBCL patients of synonymous mutations in the coding regions of the BCL2 gene, using large-scale data. In this research we analyzed large scale data in order to identify synonymous mutations with vast influence on DLBCL, and to understand their mechanisms and evolution.

2 Methods

2.1 Overview

In order to find cancerous mutations and their matching regulatory mechanisms, we have created a pipeline. As a proof of concept, we focus on synonymous mutations in the coding regions of the BCL2 gene, found in DLBCL patients, which affect the binding of the transcription factors. Transcription factors are proteins which bind to DNA sequences and affect the transcription of a certain gene – by triggering the transcription (activators) or restraining it (repressors). The overall method that was used is illustrated in Fig. 1 and includes three main steps. Step A is the mutations screening, in which we looked for synonymous mutations that were found in large number of patients. The fact that these synonymous mutations were found in large number of patients suggests that there is a selection in DLBCL towards these mutations. Stages B and C are independent. Stage B includes the phenotype understanding, in which we aim to understand the effect of the mutation. Stage C is the mechanism analysis, in which we aim to understand the mechanism behind the effect of the mutation.

2.2 Mutations Screening

Data of malignant Lymphoma patients (Germinal center B-cell derived lymphomas, GCB) were gathered from the ICGC-MMML-Seq (Molecular Mechanisms in Malignant Lymphoma by Sequencing) Consortium [15]. The data contain simple somatic mutations (SSM) of 241 patients and sequence-based gene expression data of 105 patients. Firstly, we preformed pre-processing step to enrich our SSM data. In this step we converted the mutations locations in the genome from GRCh37 to GRCh38 using UCSC genome lift tool [16] and validated the type of the mutations in the gene. Next, we filtered out mutations that were not mapped to the BCL2 gene, and carried out the following steps, as seen in Fig. 1 block A:

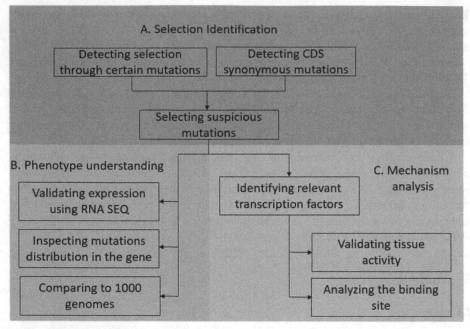

Fig. 1. A visualization of our method for finding synonymous mutations in cancer which affect the binding of transcription factors.

a. We tested which of the mutations were synonymous, i.e. do not affect the protein (amino acid) sequence. To this end, we examined the effect of the mutations on the protein sequence in every one of the possible proteins created by this gene.
b. We checked which mutations repeated themselves numerously in different patients, and in that manner detected selection in cancer in favour of these mutations. This step was done by observing only locations in which the number of mutations has a z-score > *3*, in comparison to other locations with mutations in the gene.
c. We searched for mutations that appear in both the results of the first and the second stages (the intersection). Theses mutation were further studied in-depth in the next steps.

2.3 Phenotype Understanding

Following the identification of the suspected mutations, we used the following tests in order to understand the effect of the mutations and its connection to DLBCL, as seen in Fig. 1 block B:

a. Test RNA expression levels of BCL2 using the RNA-seq data from the ICGC, to understand if the mutations influence gene expression levels. We took 10,000 random groups of eight patients and checked their average BCL2 expression, creating a distribution of expression. Then we compared it to the group of eight patients who have our suspected mutations.

b. Examine the gene's mutations distribution – in order to verify that there is a significant selection on these mutations, hence the plausibility that they have influence. To this end we created a visualization of the mutations across the gene.
c. Test if the mutations appear in 1000 genomes database [17] – the database contains the genomic sequencing of random people representing the general population. Thus if the mutation does not appear in it, but appears among cancer patients, we assume that it is associated with the disease.

2.4 Mechanism Analysis

We analysed the possible related mechanism, hence finding transcription factors that may be affected by these mutations, as seen in Fig. 1 block C. This stage starts in identification of relevant transcription factors that, according to their target sequence, will be significantly influenced by those mutations. Next, we used the transcriptions factors from the previous steps and preformed two more analysis steps:

a. Validate the presence of these transcription factors in the relevant tissue - lymphocytes. To this end we used GTEx database of expression in different tissues [18] and checked whether or not our suspected transcription factors are expressed in lymphocytes in comparison to others tissues.
b. Analyse the suggested binding site. The step includes the test of the binding strength using JASPAR's [19] PSSM binding data and the search of nearby alternative binding sites. In order to do so we looked for binding sites with same or higher binding score in proximity to our suspected binding site.

3 Results

3.1 Mutations Screening

First, we acquired all synonymous mutations of GCB lymphoma patients. Afterwards we counted the number of mutations found per location in the BCL2 gene. Lastly, we calculated the z-score for each position with synonymous mutations, referring to every position with synonymous mutations.

Out of 3266 mutations along the gene, three location were found with a z-score higher than three:

a. 18:63318600 substitution from G to A found in 9 patients.
b. 18:63318601 substitution from C to T found in 14 patients.
c. 18:63318643 substitution from C to A/T/G found in 9 patients.

Since the first two mutations are adjacent, we have hypothesized that they affect the same regulatory factor. Therefore, we decided to focus on the first two mutations, as one regulatory zone that might affect a possible transcription factor. These mutations will be referred as the 600 and 601 mutations from here on out.

3.2 Phenotype Understanding

First, we tested if mutations 600 and 601 exist in the general population, via the 1000 genomes database [17]. These mutations do not exist in 1000 genomes, which strengthens the proposition that there is connection between these mutations and cancer, specifically GCB.

We then tested the mutation dispersion and the types of mutations along the gene. As can be seen in Fig. 2, the gene is found on the complementary strand and contains at its center a large intron. There is a large concentration of mutations on the first exon, and mutations 600 and 601 are the mutations with the highest number of repeats within the coding region. This representation of GCB patients' mutations in the BCL2 gene emphasizes the selection towards these mutations in this cancer.

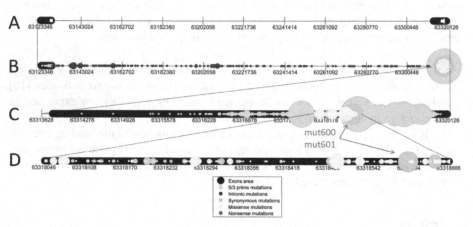

Fig. 2. This figure is divided into four parts: part A - the BCL2 gene and its coordinates in chromosome 18. The exons are marked in wide black, stop codons are presented as white boxes and the start codon is presented by white triangle. In part B we added the mutations that we found in GCB patients. Every circle portrayed a mutation found in this location, while the color represents the type of the mutation, and the size of the circle represents in how many patients it occurs. Part C is focused on the first exon, and part D is focused on the coding region of the first exon. The circle that represents mutations 600 and 601 is highlighted with green arrows (due to the fact that they are adjected, their circles look as one circle in this graph). (Color figure online)

In the next stage, we wanted to understand how these mutations influence BCL2 and what phenotype they create. For this purpose, we utilized RNA-Seq of GCB patients available in ICGC. Note that out of 241 GCB patients who have whole genome sequencing, 109 also have RNA-Seq measurements, and only 8 out of the latter group have one of the suspect mutations (600 or 601). One can see in Fig. 3 that among the patients that have one of the suspect mutations there is overexpression of BCL2, and the result is statistically significant ($p = 0.0154$) according to empirical p-value based on resampling.

Therefore, it appears that unlike the general population (as portrayed in 1000genomes), there is selection among GCB patients for the suspect synonymous mutations (600 and 601), and these mutations cause the overexpression of BCL2 which

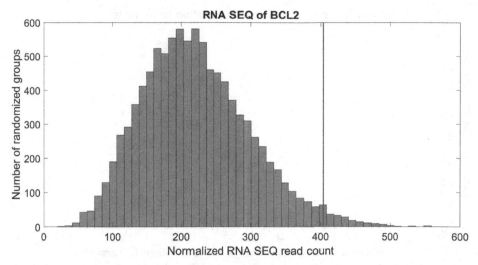

Fig. 3. We randomly sampled 10,000 times a group of 8 GCB patients, and we measured the average expression level of BCL2 for each group. This histogram represents all the groups' results, where the red line depicts the average BCL2 expression level in the group that contains all 8 patients with one of the suspect mutations (600 or 601). (Color figure online)

prevents apoptosis, and in this manner aids cancer by creating cells that cannot die via apoptosis.

3.3 Mechanism Analysis

To better understand the mechanism of regulation of these mutations we first took all transcription factors found on JASPAR [19], and calculated their maximum binding score using a sliding window over all windows in the gene containing the 600 and 601 mutation locations. Then we inserted (separately) the 600 and 601 mutations and re-calculated the binding score of each transcription factor in the same genomic area (including the mutation). Afterwards, we multiplied the given score of each mutation by the number of times the latter occurred in order to combine the two aspects. We then added the results related to each of the mutations to estimate the combined effect. The result represents how the mutations in 600 and 601 locations influenced the ability of the transcription factor to bind to the genome in this area. The results showed that most of the transcription factors are not significantly affected by these mutations, as can be seen in Fig. 4. We took all transcription factors with z-score higher than 3 or lower than -3 for further investigation.

We believe that mutations 600 and 601 cause overexpression of the BCL2 gene, and therefore if the mechanism is connected to a specific transcription factor, we will expect to find one of two options:

1. The mutations cause better binding of an activator transcription factor
2. The mutations cause worse binding of a repressor transcription factor

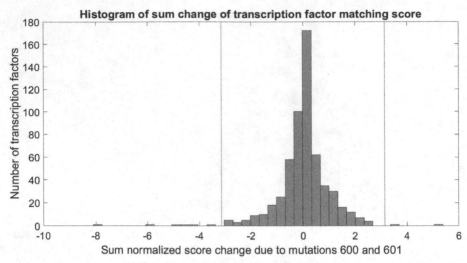

Fig. 4. Histogram representing how the different transcription factors were influenced by the 600 and 601 mutations. The X axis was calculated by multiplying the change in the transcription factor matching score by the number of times this mutation occurs, and then adding the two results together. Z-scores 3 and -3 are marked in red lines (Color figure online)

In addition, due to the fact that our patients have GCB lymphoma, we expect to see this transcription factor active in the relevant tissue – lymphocytes. In order to check the latter, we used GTEx data [18] of expression in different tissues.

We used the two methods mentioned above (mechanism logic and transcription factors expression data) for finding the relevant transcription factor out of the ones that their binding score was changed significantly. As can be seen in Fig. 5, only one transcription factor, musculin (MSC), passed all the tests. The latter binding score was changed significantly due to the mutations (z-score of -5.5231), it is a repressor that its binding score declined (-5.74391), and it is highly expressed in lymphocytes (median TPM of 143.53, 9.6801 times the mean median TPM across all tissues). MSC also known as ABF-1, and it is a well-known factor expresses in lymphocytes [20]. Methylation of MSC was shown to be correlated with DLBCL [21], as well as SNPs in the MSC gene itself [22].

Finally, in order to validate that the proposed mechanism (damage to the binding of the repressor MSC) is indeed the mechanism by which the synonymous mutations affect the expression of BCL2, we tested if the repressor attached to this area prior to the mutations.

We observed the LOGO which represents the binding site necessary for MSC. As seen in Fig. 6, prior to the mutations, the sequence in this region matches the LOGO well, with emphasis on full compatibility at the center of the region. However, mutations 600 and 601 change the sequence to completely incompatible letters (appear with probability 0 in the PSSM).

To examine if the adaptability of the region prior to the mutations is extraordinary in its quality. We examined all possible binding sites of the BCL2 gene whose binding

Transcription factor	Sum change in binding score due to 600 and 601 mutations	Sum change in binding score due to 600 and 601 mutations z-score	Transcription factor type	Median TPM in the tissue 'Cells - EBV-transform ed lymphocyt es'	Median TPM in the tissue 'Cells - EBV-transformed lymphocytes' / mean median TPM in all tissues
RHOXF1	-7.93801	-7.6267	Unknwon[24–26]	0.558	0.178
MSC	-5.74391	-5.5231	Repressor[25,26]	143.53	9.6801
FIGLA	-4.75537	-4.5753	Activator[25,26]	0.0066	0.0354
CRX	-4.50247	-4.3329	Activator[25,26]	0.01	0.7279
NR2E3	-4.05709	-3.9058	Activator[25,26]	0.024	0.069
NHLH1	-3.52741	-3.398	Activator[25,26]	1.53	2.398
FOXD2	5.367526	5.13	Activator[25,26]	0.74	0.5365
SOX17	3.63272	3.4667	Activator[25,26]	0.033	0.0023

Fig. 5. This table shows the results of the different tests done on the transcription factors with very high/low z-score due to significantly change in the matching score in 600 and 601 mutations area. The tests include the sum change in the binding score due to the mutations, the type of the transcription factor, the median TPM in lymphocytes, and the median TPM lymphocytes / mean median TPM in all tissues. In the center part of the table (columns 2–4) we highlighted the transcription factors whose type (activator/repressor) matches the change in the binding score in order to match the BCL2 overexpression. In the right part (columns 5–6) we highlighted the transcription factors that are highly expressed in lymphocytes (their expression in lymphocytes is higher than their mean median expression in all tissues).

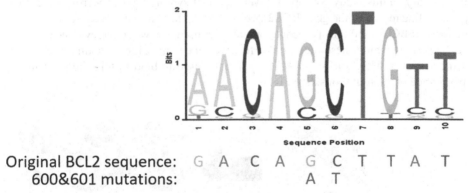

Original BCL2 sequence: G A C A G C T T A T
600&601 mutations: A T

Fig. 6. The upper part contains the LOGO of the binding site of MSC. Below you can see two rows, the first represents the original BCL2 sequence, as found in the reference genome, while the second line shows the effect of the 600 and 601 mutations on the genome.

score according to the PSSM is equal or better than our suspect site. Out of 196,783 possible double windows (for both strands) in the gene, there are only 103 site windows whose compatibility is better than our suspect site ($p = 5.2 \cdot 10^{-4}$). We then looked at the possible window mapping in the gene and discovered that our suspect site is the first site in the regular strand, and there is no better site in this strand near the beginning of the gene, as seen in Fig. 7. Thus, it appears that this is a suitable site for MSC repressor binding, and there are no superior alternative sites near the beginning of the gene on this strand, hence damaging it will prevent repressor binding in that region.

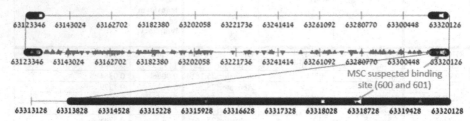

Fig. 7. A visualization of the BCL2 gene as seen in Fig. 2 (exons are in wide black, start codon in white triangle and stop codon in white box). Every red triangle represents a binding site with equal or better binding score than our suspected site, which presented as green triangle. The size of the triangle represents how good the site is for binding, according to the PSSM. The orientation of the colored triangles represents the strand on which the binding site is – normal triangle is the reverse strand and upside-down triangle is the standard strand. The third visualization in this figure is a zoom in on the first exon of the gene. (Color figure online)

4   Conclusion

In summary, in this study we report novel recurrent synonymous cancerous mutations in the coding region of the gene BCL2. We found that these mutations cause the overexpression of BCL2, and thus prevent apoptosis. Furthermore, we found evidence that the mechanism by which these mutations affect the gene expression is binding prevention of the repressor MSC to BCL2. MSC apparently used to bind to this site prior to the appearance of the mutations. See Fig. 8.

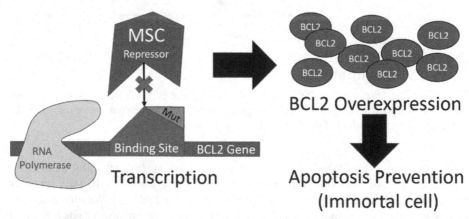

Fig. 8. An illustration of the suggested mechanism. On the left you can see an illustration of the BCL2 gene with the desired binding site of the MSC repressor. The latter cannot bind to the gene due to the mutations in the binding site. As a result, the RNA polymerase efficiently transcribes the gene, resulting in the overexpression of BCL2. The overexpression of the BCL2 gene will eventually prevent the apoptosis process which prevents the death of the cell.

We are now working on implementing the pipeline demonstrated in this study to find additional silent mutations that interfere with binding of transcription factors of other type of cancer. The full pipeline code will be published with the final genome-wide and cancer-wide study.

References

1. Futreal, P.A., Coin, L., Marshall, M., Down, T., Hubbard, T., Wooster, R., et al.: A census of human cancer genes. Nat. Rev. Cancer **1**(4), 177 (2004)
2. Bergman, S., Tuller, T.: Widespread non-modular overlapping codes in the coding regions. Phys. Biol. **17**(3), 031002 (2020)
3. Hunt, R.C., Simhadri, V.L., Iandoli, M., Sauna, Z.E., Kimchi-Sarfaty, C.: Exposing synonymous mutations. Trends Genet. **30**(7), 308–321 (2014)
4. Supek, F.: The code of silence: widespread associations between synonymous codon biases and gene function. J. Mol. Evol. **82**(1), 65–73 (2015). https://doi.org/10.1007/s00239-015-9714-8
5. Diederichs, S., Bartsch, L., Berkmann, J.C., Fröse, K., Heitmann, J., Hoppe, C., et al.: The dark matter of the cancer genome: aberrations in regulatory elements, untranslated regions, splice sites, non-coding RNA and synonymous mutations. EMBO Mol. Med. **8**(5), 442 (2016)
6. Gartner, J.J., Parker, S.C.J., Prickett, T.D., Dutton-Regester, K., Stitzel, M.L., Lin, J.C., et al.: Whole-genome sequencing identifies a recurrent functional synonymous mutation in melanoma. Proc. Natl. Acad. Sci. **110**(33), 13481 (2013)
7. Plotkin, J.B., Kudla, G.: Synonymous but not the same: the causes and consequences of codon bias. Nat. Rev. Genet. **23**(12), 32 (2010)
8. Hansen, T.V.O., Steffensen, A.Y., Jønson, L., Andersen, M.K., Ejlertsen, B., Nielsen, F.C.: The silent mutation nucleotide 744 G → A, Lys172Lys, in exon 6 of BRCA2 results in exon skipping. Breast Cancer Res. Treat. **119**(3), 547–550 (2010)
9. Liu, Y., Walavalkar, N.M., Dozmorov, M.G., Rich, S.S., Civelek, M., Guertin, M.J.: Identification of breast cancer associated variants that modulate transcription factor binding. PLoS Genet. **13**(9), e1006761 (2017)
10. Tate JG, Bamford S, Jubb HC, Sondka Z, Beare DM, Bindal N, et al.: COSMIC: the catalogue of somatic mutations in cancer. Nucleic Acids Res. **29**, p. gky1015 (2018)
11. Cory, S., Adams, J.M.: The Bcl2 family: regulators of the cellular life-or-death switch. Nat. Rev. Cancer **1**(2), 647 (2002)
12. Vaux, D.L., Cory, S., Adams, J.M.: Bcl-2 gene promotes haemopoietic cell survival and cooperates with c-myc to immortalize pre-B cells. Nature **29**(335), 440 (1988)
13. Monni, O., Franssila, K., Joensuu, H., Knuutila, S.: BCL2 overexpression in diffuse large B-cell lymphoma. Leuk. Lymphoma **34**(1–2), 45–52 (1999)
14. Lohr, J.G., Stojanov, P., Lawrence, M.S., Auclair, D., Chapuy, B., Sougnez, C., et al.: Discovery and prioritization of somatic mutations in diffuse large B-cell lymphoma (DLBCL) by whole-exome sequencing. Proc. Natl. Acad. Sci. **109**(10), 3879 (2012)
15. Zhang, J., Bajari, R., Andric, D., Gerthoffert, F., Lepsa, A., Nahal-Bose, H., et al.: The international cancer genome consortium data portal. Nat. Biotechnol. **37**(4), 367–369 (2019)
16. Kent, W.J., Sugnet, C.W., Furey, T.S., Roskin, K.M., Pringle, T.H., Zahler, A.M., et al.: The human genome browser at UCSC. Genome Res. **12**(6), 996–1006 (2002)
17. Auton, A., Abecasis, G.R., Altshuler, D.M., Durbin, R.M., Abecasis, G.R., Bentley, D.R., et al.: A global reference for human genetic variation. Nature **526**(7571), 68–74 (2015)
18. Lonsdale, J., Thomas, J., Salvatore, M., Phillips, R., Lo, E., Shad, S., et al.: The genotype-tissue expression (GTEx) project. Nat. Genet. **45**(6), 580–585 (2013)

19. Mathelier, A., Zhao, X., Zhang, A.W., Parcy, F., Worsley-Hunt, R., Arenillas, D.J., et al.: JASPAR 2014: an extensively expanded and updated open-access database of transcription factor binding profiles. Nucleic Acids Res. **42**(D1), D142–D147 (2013)
20. Massari, M.E., Rivera, R.R., Voland, J.R., Quong, M.W., Breit, T.M., van Dongen, J.J.M., et al.: Characterization of ABF-1, a novel basic helix-loop-helix transcription factor expressed in activated B lymphocytes. Mol. Cell. Biol. **18**(6), 3130 (1998)
21. Ushmorov, A., Leithäuser, F., Ritz, O., Barth, T.F.E., Möller, P., Wirth, T.: ABF-1 is frequently silenced by promoter methylation in follicular lymphoma, diffuse large B-cell lymphoma and Burkitt's lymphoma. Leukemia **27**(22), 1942 (2008)
22. Ghesquieres, H., Slager, S.L., Jardin, F., Veron, A.S., Novak, A., Maurer. M.J., et al.: A genome-wide association study (GWAS) of event-free survival. In: Diffuse Large B-Cell Lymphoma (DLBCL) Treated With Rituximab and Anthracycline-Based Chemotherapy: A Lysa and Iowa/Mayo Clinic SPORE Multistage Study. Blood, 15 November 2013, vol. 122, no. 21, p. 76 (2013)
23. Lee, S.-E., Lee, S.-Y., Lee, K.-A.: Rhox in mammalian reproduction and development. Clin. Exp. Reprod. Med. **40**(3), 107–114 (2013)
24. Ashburner, M., Ball, C.A., Blake, J.A., Botstein, D., Butler, H., Cherry, J.M., et al.: Gene Ontology: tool for the unification of biology. Nat. Genet. **25**(1), 25–29 (2000)
25. The Gene Ontology Consortium: The gene ontology resource: 20 years and still going strong. Nucleic Acids Res. **47**(D1), D330–D338 (2018)

Author Index

Printed in the United States
By Bookmasters